Viviane Vaz de Queiroz
Luciano Bem Bianchetti

The Subtribe Oncidiinae (Orchidaceae), Brazil

Viviane Vaz de Queiroz
Luciano Bem Bianchetti

The Subtribe Oncidiinae (Orchidaceae), Brazil

The Subtribe Oncidiinae Benth. (Orchidaceae Juss.) in the Federal District, Brazil

Imprint
Any brand names and product names mentioned in this book are subject to trademark, brand or patent protection and are trademarks or registered trademarks of their respective holders. The use of brand names, product names, common names, trade names, product descriptions etc. even without a particular marking in this work is in no way to be construed to mean that such names may be regarded as unrestricted in respect of trademark and brand protection legislation and could thus be used by anyone.

Cover image: www.ingimage.com

This book is a translation from the original published under ISBN 978-613-9-65558-8.

Publisher:
Sciencia Scripts
is a trademark of
Dodo Books Indian Ocean Ltd. and OmniScriptum S.R.L publishing group

120 High Road, East Finchley, London, N2 9ED, United Kingdom
Str. Armeneasca 28/1, office 1, Chisinau MD-2012, Republic of Moldova, Europe
Printed at: see last page
ISBN: 978-620-7-78405-9

SUMMARY

ACKNOWLEDGMENTS

We thank God for the gift of life and our family members who are by our side.

Thanks to the author, Professor Dr. Carolyn Elinore Barnes Proença. A woman who encourages us to grow and to be good and kind in the profession of teaching and botany.

SUMMARY

Orchids belong to one of the largest and most diverse families of flowering plants in the world. The subtribe Oncidiinae Benth. (Cymbidieae, Epidendroideae) is the second largest orchid subtribe in the Americas. According to the List of Species of the Flora of Brazil, it represents 15 genera and 19 species in the Federal District. The aim of this work was to develop a taxonomic treatment for this subtribe for the "Flora of the Federal District, Brazil". Field trips were made during the flowering season and vegetative plants were grown. The BHCB, CEN, HEPH, IBGE and UB herbaria were consulted, the latter being the institution chosen to deposit the new collections. A morphological analysis was carried out on all the material (116 collections), and a further genus and two species were found for the Federal District, *Cohniella jonesiana* (Rchb.f.) Christenson and *Trichopilia brasiliensis* Cong. The occurrence of *Coppensia bifolia* (Sims) Dumort due to an erroneously identified sample was suppressed, as was the occurrence of *Notylia hemitricha* Barb. Rodr., and *Rodriguezia brachystachys* Rchb.f. & Warm, because there is no sample or collection in herbaria that morphologically fits the description of the species and the HB herbarium that mentions the sample is closed. Through the studies carried out, the list of genera and species of Oncidiinae Benth. occurring in the Federal District was corrected to 15 genera and 18 species, namely: *Alatiglossum fuscopetalum* (Hoehne) Baptista, *Alatiglossum macropetalum* (Lindl.) Baptista, *Aspasia variegata* Lindl., *Cohniella cepula* (Hoffmanns.) Carnevali & G Romero, *Cohniella jonesiana* (Rchb.f.) Christenson, *Comparettia coccinea* Lindl, *Coppensia hydrophila* (Barb.Rodr.) Campacci, *Coppensia varicosa* (Lindl.) Campacci, *Ionopsis utricularoides* (Sw.) Lindl., *Lockhartia goyazensis* Rchb.f., *Lophiaris pumila* (Lindl.) Braem, *Macroclinium Wullschlaegelium* Focke, *Notylia lyrata* S.Moore, *Plectrophora edwallii* Cogn., *Rodriguezia decora* (Lem.) Rchb.f., *Sanderella discolor* (Barb.Rodr.) Cogn., *Trichocentrum albococcineum* Lindl., and *Trichopilia brasiliensis* Cong. All the genera and species were described and identification keys and illustrations were drawn up for the genera and species. It was concluded that some species of Oncidiinae occurring in the Federal District are widely distributed, while others are locally rare and require intense collection efforts to be recorded.

1. INTRODUCTION

1.1. THE ORCHIDACEAE FAMILY

Orchids are monocots and belong to one of the largest and most diverse families of flowering plants in the world (Dressler, 1981). The group has evolved highly specialized adaptations to attract, deceive and manipulate insects for pollination, and this has fascinated researchers and observers since Darwin's time (Dressler, 1981). Many species offer nectar or floral oils as a reward to pollinators, however, most species do not offer any type of reward and, for this reason, these species achieve pollination through "deception" strategies, with their flowers mimicking the shape, emitting characteristic odors or sharing environments with species (belonging to other families, such as Malpighiaceae) that offer a reward (Chase *et al.*, 2009).

Orchids vary in floral and vegetative morphology, ranging from plants with millimetre-sized flowers (e.g. *Lophiaris pumila* (Lindl.) Braem) to plants larger than a person (e.g. *Grammatophyllum speciosum* Blume), and flowers with very attractive colors to neutral colors and pleasant to unpleasant scents (Baptista *et al.*, 2011).

It is estimated that there are around 25,400 species in the world, divided into around 1,000 genera (Baptista *et al.*, 2011). Representatives of the Orchidaceae family are distributed in almost all parts of the world, except in the extremely arid polar and desert regions (Baptista *et al.*, 2011). They are strongly represented in the Americas, with Colombia having the largest number, with 4,010 species, followed by Ecuador, with 3,784 species and Brazil, with 2,459 species (Dodson, 2003; Sarmiento, 2007; Barros *et al.*, 2015).

The ability of orchids to adapt to various types of vegetation is due, among other things, to the different vegetative structures present in the family, presenting strategies for capturing and reserving water and nutrients, pseudobulb-like stems and fleshy leaves with the function of storing reserves, velamen-like roots with the function of rapid absorption of water and nutrients and their own growth in clumps, allowing the accumulation of organic matter (Hoehne, 1949).

Orchids are herbaceous perennial plants and their behavior can vary between epiphytes (around 70%), terrestrial, rupestrian, climbers and saprophytes (Silva, 1999).

Orchidaceae are characterized by having fasciculated roots which, in some species, are fleshy and swollen (tuberoid). In the vast majority of species, especially epiphytes, the roots are coated with a multiseriate, spongy, whitish epidermis called a velamen, which has the function of rapidly absorbing water and protecting against drying out. Secondary stems can be swollen into pseudobulbs to store water and nutrients. The leaves are alternate, distichous or spiral. Prefoliation can be duplicate, conduplicate, plicate or convolute. The flowers are hermaphrodite, or rarely unisexual (dimorphic, in *Catasetum* Rich. *ex* Kunth), often zygomorphic, rarely asymmetrical, usually trimerous, with one of the petals opposite the fertile stamen, morphologically modified, forming the lip. The lip can be concrescent with the column and the basal and central regions are usually calloused. The callus can be in the form of a keel, a plate, or have cornicles, warts, oil-secreting structures or osmophores, which emit odors. The androecium is made up of one, rarely two or three fertile stamens. The fillet is attached to the stipe forming the column or gynostemium; the stigma is usually located on the ventral side of the gynostemium, is three-lobed, with one of the lobes being partially sterile, forming the rostellum, which separates the anther from the stigma. The anther, in most cases, is represented by a "hood" to protect the pollen and is usually deciduous when the pollen is removed. The pollen, in most species, is united in pollinia, in numbers of 2, 4, 6 or 8. The ovary is infertile, unilocular, tricarpellar, with parietal placentation. The fruits are capsular, usually dry, although in some genera of epiphytes they can appear as fleshy capsules. The seeds are numerous, tiny, with a rudimentary embryo, devoid of endosperm, adapted to dissemination by the wind, an important fact for the epiphytic life form (Weberling & Schwantes, 1986; Silva, 1999; Chase, 2009; Baptista *et al.*, 2011; Rodrigues, 2011).

After phylogeny studies using molecular data, and according to the type of aggregation of the pollen grains (characteristics of the pollinas), anthers and leaves, the Orchidaceae family was divided into five subfamilies: Apostasioideae, Cypripedioideae, Epidendroideae, Orchidoideae and Vanillioideae (figure 1) (Pridgeon *et al.*, 1999).

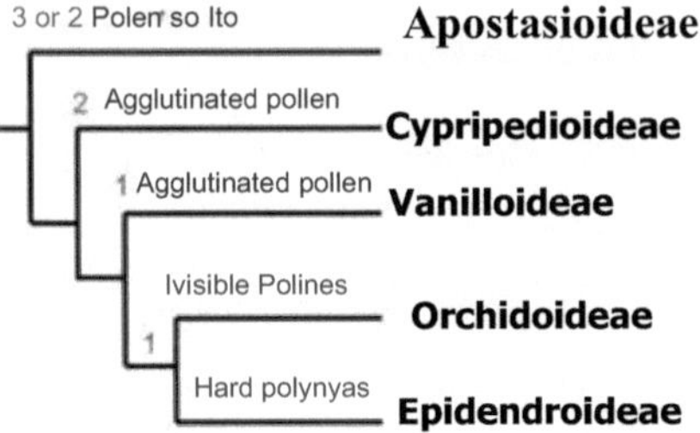

(*) Number of fertile anthers

(*) and (*) Type of pollen aggregation and consistency

The subfamily to be demonstrated in this book, Epidendroideae Lindley, has almost 600 genera (Baptista *et al.*, 2011). It has a large number of epiphytic species, and is characterized by having terminal anthers, 2-8 rigid pollinia, with a cartilaginous consistency, usually with appendages such as caudicle, stipe and viscidium (Baptista *et al.*, 2011). The gynostemium can extend to form a foot of the column, with the base of the lateral sepals attached to it forming a mentum (Rodrigues, 2011). This subfamily is divided into 16 different tribes (Arethuseae, Calypsoeae, Collabieae, Cymbidieae, Dendrobieae, Epidendreae, Gastrodieae, Malaxideae, Neottieae, Nervilieae, Podochileae, Sobralieae, Triphoreae, Tropidieae, Vandeae, Xerorchideae) (Baptista *et al.*, 2011).

The tribe Cymbidieae Pfitzer has the largest number of species, with distribution on all continents and possibly including a third of Brazilian species (Baptista *et al.*, 2011). Morphologically, the tribe Cymbidieae Pfitzer is generally characterized by medium-sized plants, sympodial growth, the presence of a velamen, pseudobulbs, articulated leaves, inflorescences usually at the base of the pseudobulbs, flowers with a stipe, and two to four pollinia (Baptista *et al.*, 2011). The tribe Cymbidieae Pfitzer has nine subtribes (Baptista *et al.*, 2011), including the subtribe Oncidiinae Benth.

1.2. THE SUBTRIBE ONCIDIINAE BENTH.

For a better understanding of the nomenclature of the species belonging to the subtribe Oncidiinae, it is necessary to briefly review the genus *Oncidium* Swartz, the type of this subtribe.

Until recently, the taxonomy of the genus *Oncidium* was particularly difficult, especially because of the criteria used to classify the species and the imprecise or poorly defined generic limits.

The genus *Oncidium* was established in 1800 by Olof Swartz and has around 520 species, most of which are epiphytes and some terrestrial (Senghas, 1998 *apud* Faria, 2004). The name *Oncidium* (Greek: *onkidion* diminutive of *onkos* = tumor, nodule, swelling), is an allusion to the small callus that occurs at the base of the lip (Chase, 2009). Originally the genus was defined based on five specimens representing five different species, namely: *Oncidium altissimum* (Jacq.) Sw. (originally described as *Epidendrum altissimum* by Jacquin in 1760); *Oncidium cebolleta* (Jacq.) Sw. (originally described as *Epidendrum cebolleta* by Jacquin in 1760); *Oncidium carthagenense* (Jacq.) Sw. (originally described as *Epidendrum carthagenense* by Jacquin in 1760); *Oncidium quadripetalum Sw.* (originally described as Epidendrum *tetrapetalum* by Jacquin in 1760) and, finally, *Oncidium variegatum (*Sw.) Sw. (originally described as *Epidendrum variegatum* by Swartz in 1788) (Braem, 1993).

At the time the genus was described, the criteria used by Swartz were sufficient to distinguish *Oncidium* species from other genera. However, as the years went by, new species were described and, due to the fact that the circumscription or generic limits were imprecise or poorly defined, all the species that "looked like" *Oncidium*, even those that showed morphological discrepancies, were "accommodated" in the genus, which included many species. In addition to the large number of species, some of the characteristics presented increased the problems related to taxonomy, such as: the species had a wide geographical distribution (Florida, USA to Argentina); diverse behaviors (epiphytic, terrestrial and rupicolous); vegetated in diverse environments (areas with high humidity to desert areas); a lot of variation in external morphology and a lot of variation in chromosome numbers (Braem, 1993).

The first revision of the genus *Oncidium* was carried out by Fritz Kraenzlin (Kraenzlin, 1922) and published in Das Pflanzenreich. This work was considered to be rather imprecise and unsatisfactory for species identification, and not very useful (Braem, 1993). Among the many problems with Kraenzlin's work, many of his illustrations did not coincide with the descriptions of the species (Braem, 1993; Faria, 2004), leading scholars to make mistakes in their identifications. For more than fifty years, after Kraenzlin's work (1922), many authors had expressed the need for a new revision of the genus, claiming that many species should be segregated from *Oncidium* and placed in autonomous genera. It wasn't until 1974 that Garay & Stacy proposed a new classification for *Oncidium*. After various considerations, the authors subdivided *Oncidium* into 26 sections based on floral characters. In the same study, Garay and Stacy (1974) also lectotypified *Oncidium variegatum* as the type species for the genus, as it was the only species of the genus illustrated by Swartz in his original publication of *Oncidium*. Later, in 1982, Dressler and Williams proposed the indication of *Oncidium altissimum* (Jacq.) Sw. as a lectotype, a proposal accepted by the *Committee for Spermatophyta* (Stafleu, 1985). This was the beginning of a new revision of the genus *Oncidium s.l.* in which all species belonging to the *O. altissimum* section would be considered the "true" *Oncidium*. All other species were to be "accommodated" in other genera and therefore given new names (Braem, 1993).

The subtribe Oncidiinae Benth. (Cymbidieae, Epidendroideae) is one of the most diverse subtribes of Orchidaceae and is the second largest subtribe in the Americas with an expressive diversity in floral and vegetative morphology, comprising 69 genera (Chase, 2009) and around 1,700 species with an exclusively Neotropical distribution (Penha *et al.*, 2011). In Brazil, the subtribe Oncidiinae Benth. is represented by around 49 genera and 360 species (Barros *et al.*, 2015). In the Federal District, it is represented by 15 genera and 19 species (Barros *et al.*, 2015).

Before molecular phylogenetic studies, the delimitation of the subtribe Oncidiinae was quite varied, from the relatively broad concept of Dressler (1993) to the relatively narrow concepts of Szlachetko (1995).

Based solely on morphology, Dressler (1993) considered the subtribes Ornithocephalinae and Telipogoninae to be close but distinct subtribes of Oncidiinae. Whitten *et al.* (2000) and Chase (2009), based on molecular studies, demonstrated that the subtribes Ornithocephalinae, Pachyphyllinae and Telipogoninae are "embedded" within Oncidiinae and this new scope for Oncidiinae is accepted by most taxonomists.

The flowers of the subtribe Oncidiinae have a great diversity of shape, size and function, involving a range of pollinators, which makes them attractive for evolutionary studies (Neubig, 2012). As far as floral rewards are concerned, their resources include nectar, oils and fragrances. Pollination strategies include deception (of flowers that don't offer rewards) and mimicking these flowers with flowers that do offer rewards (especially those of the Malpighiaceae family) (Neubig, 2012). In this group, the number of chromosomes is also quite variable, ranging from the smallest known number in orchids, $2n = 10$, to $2n = 168$ (Felix and Guerra, 2000; Neubig, 2012).

Some work proposed for Orchidaceae (Chase & Palmer, 1989) and, more specifically for Oncidiinae Benth. (Chase *et al.*, 2009), suggests that the use of floral characteristics alone should not be used to define generic boundaries and formulate phylogenetic hypotheses.

Phylogenetic studies have shown that many genera of Oncidiinae Benth. in their traditional circumscription are polyphyletic or paraphyletic (Williams *et al.*, 2001). From a modern perspective, in order to establish the monophyly of the genera, it is necessary to expand the circumscription of some genera, along with the necessary nomenclatural changes (Williams *et al.*, 2001).

Some of the new phylogenetic proposals (Chase *et al.*, 2009) for the genera of Oncidiinae Benth., in an attempt to make monophyletic arrangements, suggest that species belonging to autonomous genera such as *Alatiglossum* D.H. Baptista, *Baptistonia* Barb. Rodr., *Binotia* Rolfe, *Carriella* V.P.Castro & K.G.Lacerda, *Coppensia* Dumort, *Ornithophora* Barb. Rodr., *Rodrigueziella* Kuntze and *Rodrigueziopsis* Schltr., among others, should be transferred to *Gomesa s.l. Similarly*, work by Williams *et al.* (2001), corroborated by Chase (2009), also suggests new arrangements, transferring species belonging to genera such as *Cohniella* Pfitzer, *Lophiarella* Szlach., *Lophiaris* Raf., among others, to *Trichocentrum s.l.* Here it should be considered that many authors (Sandoval-Zapotitla & Terrazas, 2001; Sosa *et al., 2001;* Williams et al., 2001; Chase, 2009; Chase *et al.,* 2009, WCSP, 2014; among others) accept the proposals made by Williams et al. *(*2001) and Chase *et al.* (2009).

Disagreeing with the new arrangements proposed above, in this book we have adopted the List of species of the Flora of Brazil (2014), based on Braem (1993), Pupulin (1995), Koniger & Pongratz (1997, 1999), Jiménez-Machorro & Carnevali (2001), Pupulin & Carnevali (2005), Carnevali *et al.* (2013). These authors recognize smaller genera (*Alatiglossum* D.H. Baptista, *Coppensia* Dumort, *Cohniella* Pfitzer, *Lophiaris* Raf. and *Trichocentrum* Poepp. & Endl. (*s.s.*), among others), as long as they have monophyletic support, justifying the recognition of morphological and anatomical characters and the ease of diagnosing genera with a narrower delimitation.

1.3. DIVERSITY AND DISTRIBUTION IN BRAZIL

Brazil has one of the greatest orchid riches on the American continent and in the world, with around 238 genera and 2,553 species, of which 1,636 are endemic, distributed in all Brazilian plant formations (Barros *et al.,* 2015).

The representativeness of the Orchidaceae family in Brazil points to the Atlantic Rainforest as one of the regions with the greatest diversity in orchids, with 186 genera and 1,577 species, followed by the Amazon Rainforest, with 159 genera and 885 species. The Cerrado is the third most representative biome for orchids, with 138 genera and 730 species (Lista de Espécies da Flora do Brasil, 2015).

In Brazil, the subtribe Oncidiinae is represented by 49 genera and 360 species (Barros *et al.,* 2015). According to the List of Species of the Flora of Brazil, the genera of Oncidiinae are very well represented in the Atlantic Forest biome and to a lesser extent in the other biomes, with low representation in the Caatinga and Pampa biomes. They are represented as follows: Atlantic Forest, 41 genera and 242 species (83.6% representativeness); Amazon, 30 genera and 120 species (59.1%); Cerrado, 28 genera and 76 species (57.1%); Caatinga nine genera and 10 species (18.3%) (Barros *et al.,* 2015).

1.4. DIVERSITY AND DISTRIBUTION IN THE CERRADO BIOME

Knowledge of the Cerrado flora has grown a lot in the last 20 years. UNESCo (2002) estimated that the Cerrado flora has more than 6,000 species of vascular plants, with many endemic species. The 2015 Flora of Brazil species list of Angiosperms from the Cerrado Biome includes 181 families, 1,673 genera and 12,384 species.

For the flora of the Cerrado biome, Orchidaceae stands out as one of the most species-rich families. From the pioneering work by Warming (1892), later translated (Warming, 1973), to recent floristic surveys in the Cerrado Biome, Orchidaceae invariably appears as one of the four most numerically representative families in a universe of around 171 families (Mendonça *et al.*, 2008). Mendonça *et al.* (2008) listed 121 genera and 666 species for the biome and, in 2014, 138 genera and 730 species were listed, as a result of new studies and the drawing up of lists for the orchid flora of Brazil (Lista de Espécies da Flora do Brasil, 2015).

1.5. DiVERSiTY AND DiSTRiBUTION IN THE FEDERAL DiSTRiT

The List of the Flora of Brazil 2015 records 156 families and approximately 1,006 genera and 3,414 species of phanerogams in the Federal District, of which 80 genera and 252 species belong to the Orchidaceae family. For the Federal District, 15 genera and 19 species belonging to the subtribe Oncidiinae were recorded.

In the Federal District, floristic and morphological studies have been carried out that have generated lists of Orchidaceae, such as: List of vascular species recorded in the IBGE Ecological Reserve, divided by major taxonomic groups and families and relating habits and environments of occurrence (Pereira *et al.*, 1993), Floristic survey of the Aguas Emendadas ecological station (Maury *et al.*, 1994), The vegetation of the Sucupira farm (Walter & Sampaio, 1998), List and protection level of phanerogam species in the Federal District (Proença *et al.*, 2001), Floristic and physiognomic survey of the Ezechias Heringer Ecological Park (Guarà Park), Federal District (Nogueira *et al.*, 2002), Updated list of Orchidaceae in the District

Federal (Batista & Bianchetti, 2003), Study of the genus *Oncidium* (Pellizzaro *et al.*, 2004), Orchidaceae of the Guarà Ecological Reserve, Federal District (Batista *et al., 2005),* Orchidaceae family in the Cafuringa APA and its contribution to the diversity of the flora of the Federal District and the Cerrado biome (Bianchetti *et al.*, 2005), The genus *Encyclia* (Orchidaceae) in the Federal District, Goiàs and Tocantins (Meneguzzo *et al., 2012).*

2. BACKGROUND

The flora of the Federal District is being studied more and more, but orchids have few generic treatments of specific groups for the Federal District region (Batista & Bianchetti, 2003).

For the Federal District, the subtribe Oncidiinae Benth. was chosen because it has around 15 genera and 19 species and lacks adequate treatment, such as the availability of an identification key and information on the species, ecological data and habitats.

The purpose of this study is to expand the survey of the Subtribe Oncidiinae Benth. in the Federal District, to refine knowledge of its location and habitats of occurrence and to provide tools (keys and species descriptions) to enable species identification.

3. METHODOLOGY

With regard to the taxonomic approach used in this work, we agree with the aforementioned authors with regard to the preference for recognizing smaller genera, with narrower delimitation, monophyletic support and easy morphological recognition, such as *Alatiglossum* D.H. Baptista and *Coppensia* Dumort, rather than *Gomesa* (*s.l.*), and *Cohniella* Pfitzer, *Lophiaris* Raf. and *Trichocentrum* Poepp. & Endl. (n.*s.*) instead of *Trichocentrum* (l.*s.*). In addition, we corroborate the same approach used by the Lista de Espécies da Flora do Brasil, an initiative that needs to be strengthened and which will serve as a pillar for all botanical research, especially that carried out in Brazil.

In order to find out more about the species of the subtribe Oncidiinae Benth. present in the Federal District, a survey was first carried out in the herbaria of Brasilia and INCT - Herbârio Virtual da Flora e dos Fungos (2014). The survey recorded label data, with collection sites, collectors, dates and relevant information. Materials deposited in the following herbaria were consulted and analyzed: Empresa Brasileira de Agricultura e Pecuâria - Recursos Genéticos e Biotecnologia (CEN); Herbarium of the Botany Department of UFMG (BHCB); Jardim Botànico de Brasilia - Herbârio Ezechias Paulo Heringer (HEPH); Reserva Ecològica do Instituto Brasileiro de Geografia e Estatistica (IBGE); Universidade de Brasilia (UB).

In sporadic cases where certain species were mentioned for the Federal District, but without *voucher* records in the herbaria visited or *vouchers* not seen (e.g. *Cohniella jonesiana, Gomesa foliosa* and *Rodriguezia brachystachys*), the descriptions were based on material collected outside the Federal District and sometimes supplemented with data from the relevant bibliography.

To prepare the descriptions of genera and species, we used a spreadsheet of important taxonomic characters for characterizing Orchidaceae and recorded the characteristics observed (Appendix I). To describe the genera, specialized bibliography was used, which included the diagnostic characteristics for each genus (Garay & Stacy, 1974; Braem, 1993; Williams *et al.,* 2001; Chase, 2009 (Genera Orchidacearum) and http://florabrasiliensis.cria.org.br/). The generic descriptions mainly cover the total morphological variability of the genus (except for *Ionopsis*) and do not go into the characteristics presented exclusively by the species occurring in the Federal District. Most of the fruits were not seen and the descriptions were taken from the relevant bibliography (Genera Orchidacearum and http://florabrasiliensis.cria.org.br/.).

Comments were made on the etymology, general distribution, distribution in the Cerrado biome and in the Federal District; where possible, data on pollination mechanisms and probable pollinators were recorded. And a key to the genera of Oncidiinae Benth. with morphological characteristics of the genus.

Data on the geographic distribution of genera and species was taken from specific works and especially from the compilation of data available on the websites (www.floradobrasil.jbrj.gov.br., www.tropicos.org. and www.apps.kew.org/wcsp.).

A total of 55 floral diagnosis sheets were made (in order to understand and measure the specimens), including all the individuals examined that did not already have these sheets attached to the *vouchers*. One flower was taken from each exsiccata, rehydrated and mounted on cardboard under a magnifying glass. When possible, the labellum was drawn, emphasizing the disc region (callus) and the column, for a better understanding of these structures and morphology.

Morphological measurements of both the vegetative and reproductive parts (except floral diagnoses) were taken with a ruler and recorded on a spreadsheet for description. The identification of the material collected and the material already deposited in herbaria, without identification, was based on the works of Garay & Stacy (1974), Silva (1999), Pellizzaro *et al.* (2004), Chase *et al.* (2009), and Baptista et *al.* (2011) as well as by comparison with material identified by specialists in the family. Whenever possible, the standard materials requested were analyzed using phototypes available in online herbaria. Requests to borrow material from other herbaria, BHCB, from Belo Horizonte and Herbârio Mello Leitao - MBML-Herbârio do Espirito Santo were sent via request from the UB herbarium, and only the material from BHCB was received.

For each species analyzed, a broad and detailed description was made, along the lines of the Flora of the Federal District Project, as well as illustrations, and information on habitat, plant occurrence and phenology was compiled from the data on the labels of the exsiccates.

To prepare the identification keys, the differential characters between genera and species were tabulated and analyzed for variability and ease of interpretation.

A bibliographical survey was carried out on the Orchidaceae, especially the subtribe Oncidiinae Benth., in scientific papers in the form of theses, dissertations, periodicals, books and scientific websites, with a view to monitoring the evolution of knowledge about the group, with an emphasis on taxonomy and phylogeny.

Field collections were also carried out. After the survey, the presence of many collections of Oncidiinae Benth. was detected, and for this reason the intensity of the collections previously mentioned in the project was reduced. As a result, nine field trips were carried out. For the field collections, pruners were used for epiphytic species located above 2 meters in height; for the others, pruning shears were used.

The first expedition, in October 2013, to the Botanical Garden of Brasilia (Mata do Cristo

Redentor), resulted in the observation of 10 species of the family. However, none of them belonged to the subtribe Oncidiinae. The second expedition was to the IBGE Reserve (Ponte Corujao, and Mata Boca d'Agua), resulting in the observation of eight species of the family and one belonging to the Oncidiinae subtribe *Notylia lyrata* S.Moore without a flower. A third expedition was carried out in Brasilia National Park, in the Córrego do Acampamento forest, where 15 species of the family were observed. Of these, two belonged to the subtribe Oncidiinae (already cited for the Federal District), *Notylia lyrata* S.Moore and *Rodriguezia decora* (Lem.) Rchb.f. Several field trips were made, to places such as Fazenda Dois Irmaos (APA DE CAFURINGA), and species belonging to the Orchidaceae were observed, but no representative of the subtribe was found. A field trip was also made to DF-100, on the border between Goià and Distrito Federal, but no species of Oncidiinae were found in the gallery forests. A field trip was made to FERCAL and a sample of *Cohniella sp. was* collected in the dry deciduous forest, without a flower, and *Lockhartia goyazensis* Rchb.f. with fruits was collected in the anthropized gallery forest. The last field trip was outside the Federal District, in Formosa, at the Salto de Itiquira, where we observed many orchids and collected *Alatiglossum sp.*, *Cohniella sp.*, *Plectrophora edwallii* Cogn. The plants that were collected without flowers were placed in a nursery for future flowering and identification. After flowering, the plants were herborized following the traditional methodology (Walter & Cavalcanti, 2005) and, after identification, the *vouchers* were added to the Herbarium of the University of Brasilia (UB) and the duplicates sent to the Herbarium of Embrapa CENARGEN (CEN).

4.   Results and discussion

With this work, another genus and two species were recorded (*Cohniella jonesiana* (Rchb. f.) Christenson. and *Trichopilia brasiliensis* Cogn.), updating the list of species from the Flora of Brazil with 15 genera and 18 species, belonging to the subtribe Oncidiinae, in the Federal District.

TAXONOMIC TREATMENT

SUBTRIBE **ONCIDIINAE** Benth. *J. Linn. Soc., Bot., 18, 288.* (1881).

Generally epiphytic **herbs**, less often terrestrial, sympodial growth. Heteroblastic **pseudobulbs**, base usually protected by developed sheaths with or without limbs, topped by 1-3 leaves. **Leaves** distichous, apical, usually conduplicate, bilateral to unilateral, sometimes cylindrical or terete or laterally flattened, usually articulated. **Inflorescence** usually lateral (rarely terminal), simple to branched, 1 to multiflowered, usually protected by a sheath. **Flowers** usually resupinate, small to large (maximum 10 cm in diameter). **Sepals** usually free, in some cases synsepalous lateral sepals and occasionally forming a spur. **Petals generally free**, rarely attached to the base of the column. **Lip usually free**, usually without nectary, sometimes with nectariferous spur, callus often producing oil, with complex morphology. **Column** erect, usually elongated, often with wings or other appendages, sometimes at the base showing tissue similar to the callus of the lip (infra-stigmatic tubule), rarely with a projected base forming a foot of the column; clinandrium usually smooth, in some cases very developed, lacerated; anther terminal, operculate; pollinarium formed by 2-4 pollinia, compact and rigid, viscidium present; ventral stigma, entire or bilobed, usually near the apex of the column; rostellum often tapered, sometimes elongated. Ovary glabrous, rarely equinate. **Capsule** 3-(6) rimosa.

KEY TO THE GENERA OF ONCIDIINAE OCCURRING IN THE

FEDERAL DISTRICT, BRAZIL

1 - Leaves equitant, not articulated *Lockhartia*

(*L. goyazensis* fig. 2. A-B)

1' - Leaves with different morphology, articulated _____________________________ 2

2 - *Cohniella* teretes leaves

2' - Leaves ensiform, flabeliform or conduplicate but never terete ____________ 3

3 - Ensiform or flabeliform leaves ________________________________ 4

3' - Leaves conduplicated in other ways ________________________________ 5

4 - Inflorescence uniflorous; flowers with spore *Plectrophora*

(*P. edwallii* fig. 2. C-D)

4' - Multiflorous inflorescence; flowers without spore *Macroclinium*

(*M. wullsclaegelianum* fig. 2. E-F)

5 - Fleshy leaves, usually spotted with purple __________________________________ 6

5' - Leathery leaves without spots 7

6 - Flowers with a spur; column not winged, with developed apical arms ___________

     Trichocentrum

(*T. albococcineum* and fig. 3. G-H)

6' - Flowers without spur; winged column, without apical arms *Lophiaris*

(*L. pumila*)

7 - Lip inserted halfway down the spine *Aspasia*

(*A. variegata* fig. 1. G-H)

7' - Lip inserted at the base of the column ________________________________ 8

8 - Lateral sepals completely fused to form *Comparettia* spur

(*C. coccinea* fig. 1. C-D)

8' - Iateral sepals free or partially fused, without a spur or forming a small mentum   9

9 - Clinandrium very developed (above the anther), lacerated *Trichopilia*

(*T. brasiliensis* fig. 3. D)

9' - Clinandrium poorly or not developed, smooth or whole 10

10 - Column with developed wings or stellidia flanking the stigma; presence of infra-stigmatic tubule 11 10' - Column without wings or stellidia; absence of infra-stigmatic tubule 12

11 - Petals wider and longer than sepals *Alatiglossum*

11' - Petals similar in size to the sepals *Coppensia*

12 - Lip obovate, oblanceolate, oblong, oval, apex usually emarginate _ 13

12' - Lip hastened, apex acute, not emarginate *Notylia*

(*N. lyrata* fig. 2. G)

13 - Lip obovate, oblanceolate; apex very dilated in relation to the base, emarginate   14

13' - Lip oblong to pandered; apex not dilated in relation to the base, not emarginated *Sanderella*

(*S. discollor*)

14 - Panicle-like inflorescence; inconspicuous pseudobulbs; short rhizomes between secondary stems *Ionopsis*

(*I. utricularioides* fig. 3. E-F)

14' - Racemose   inflorescence; conspicuous pseudobulbs; short to extremely long rhizomes between the secondary stems *Rodriguezia*

DESCRIPTIONS OF GENERA AND SPECIES

1. *Alatiglossum* Baptista, *Colet. Orq. Bras.* 2006.

Epiphytic **herbs.** Short **rhizome** between secondary stems. Secondary stems swollen into **pseudobulbs**, fusiform to oval, sometimes tetragonal, usually furrowed and ancipitate, partially covered by leaf sheaths, topped by 1-2 leaves. **Leaves** conduplicate, bifacial, leathery, articulated, green. Lateral **inflorescence,** raceme or panicle type, longer than the leaves, multiflora. **Flowers are** resupinate, spreading and showy, usually yellow and occasionally with brown or red markings. Dorsal sepal free; lateral sepals partially fused. Free **petals**, usually larger, wider and of a different color to the sepals. **Lip** trilobed, echolobed, base of lip inserted perpendicularly to base of column, lateral lobes auriculate; region of disc with rather verrucous callus; margins of disc and isthmus usually toothed or fimbriated. **Column** with semi-circular or stellate wings flanking the stigma; presence of a prominent infra-stigmatic tubule at the base, usually composed of two lamellae, occasionally a groove; clinandrium poorly developed, smooth. **Fruit** cylindrical-fusiform capsule, not to slightly 3-angled; mature fruit with apex of carpels united.

A genus established by the Brazilian botanist Dalton Holland Baptista in 2006, alluding to the wing-like lateral lobes of the lip (Docha Neto, 2006).

The genus has around 16 species, distributed in Argentina, Bolivia and Brazil (Docha Neto, 2006). In Brazil, there are 14 species (11 endemic) in the Atlantic Forest, Cerrado and semi-arid regions of the Northeast (Docha Neto, 2006). Five species occur in the Cerrado biome and of these, two occur in the Federal District (*Alatiglossum fuscopetalum* (Hoehne) Baptista and *Alatiglossum macropetalum* (Lindl.) Baptista) (Barros *et al.,* 2015).

There is no information on pollination mechanisms or pollinators for *Alatiglossum* species. However, the floral morphology, especially the callus morphology, indicates that they belong to the group of Oncidiinae with flowers that produce oils and are probably pollinated by oil-collecting bees, similar to certain species of *Oncidium* that exhibit this syndrome.

KEY TO *ALATIGLOSSUM* SPECIES IN THE FEDERAL DISTRICT

1. Petals similar to sepals, obtuse apex, yellow and densely spotted to completely brown   1. *A. fuscopetalum*

1. Petals larger than sepals, emarginate apex, completely yellow or spotted with brown only at the base 2. *A. macropetalum*

1.1 *Alatiglossum fuscopetalum* (Hoehne) Baptista. Collect. Orquideas Brasil. 3: 88 (2006).

Fig. 4. A

Greenish-yellow **pseudobulbs**, 2.0-5.0 x 0.7-1.8 cm, persistent, articulated sheaths, 6.8-7.5 cm long, topped by 1 leaf. **Leaves** sessile, elliptic, base attenuated, apex acute, 7.0-17.7 x 0.7-1.6 cm. Paniculate **inflorescence**, 4-11 flowers, 30.0-64.0 cm long overall. **Flowers** pedicellate, pedicels 0.8-2.1 cm long. **Sepals** similar, bases attenuate, margins sinuate, apices obtuse to acuminate, membranaceous, yellowish-brown, densely spotted with dark brown, dorsal sepal obovate, 0.6-0.9 x 0.2-0.5 cm, lateral sepals 1\3 adnate, elliptic, 0.8-1.3 x 0.2-0.4 cm. **Petals** obovate to broadly elliptic, base slightly angled, margin slightly sinuate, apex obtuse, mucronate, membranous, densely to completely spotted with dark brown, 0.8-1.1 x 0.6-0.8 cm. **Lip** yellow with the central region of the callus dotted with light brown, 1.4-1.8 x 1.2-1.8 cm; lateral lobes elliptical to oval, yellow, 2.0-4.0 x 4.0-5.0 mm; median lobe oval, margin slightly erose, emarginate, 0.8-1.1 x 1.2-1.8 cm; isthmus 4.0 mm, yellow; disc region with complex callus, basal part formed by plate with fused verrucous protuberances, median part formed by bifid lamella, lower part of lamella with lateral protuberances. **Column** 2.0-3.0 x 1.0 mm, brown, infra-stigmatic tabula 1.0 x 1.0 mm, wings or stellidia yellow 2.0 x 1.0 mm. **Fruit** not seen.

Alatiglossum fuscopetalum occurs in Bolivia, Paraguay and Brazil (Tropicos, 2014; WCSP, 2014). In Brazil, it is distributed in the Federal District and in the states of Goiàs, Maranhao, Mato Grosso, Parà and Tocantins, in the Midwest, Northeast and North regions and in the Amazon and Cerrado biomes. It grows in Ciliary Forest or Gallery Forest, Dry Deciduous Forest and Dry Semideciduous Forest (Barros *et al.*, 2015). In the Federal District, it grows in mesophytic forests (Dry Deciduous Forest) with limestone outcrops and flowers in July and August.

Selected material: **APA de Cafuringa**, 15°33'S, 48°06'O, VIII.2002, *Miranda, Z.J.G. 82* (UB). **FERCAL**, 6.7 km from the CIPLAN cement factory, VII.1990, *Bianchetti, L. 869-B* (CEN).

Additional material examined: GOIAS. **Niquelândia,** Future reservoir of AHE Serra da Mesa, VII.1997, *Walter, B.M.T. 3774* (CEN).

1.2 *Alatiglossum macropetalum* (Lindl.) Baptista. Collect. Orquideas Brasil. 3: 88. (2006).

Fig. 4. A

Greenish-yellow **pseudobulbs**, 1.5-4.5 x 0.7-2.1 cm, persistent, articulated sheaths, 2.0-9.7 cm long, topped by 1 leaf. **Leaves** sessile, elliptic, apex acute, base attenuated, 2.2-17.5 x 0.5-1.8 cm. Paniculate **inflorescence**, 8-50 flowers, 27.0-76.7 cm long overall. **Flowers** pedicellate, pedicels 1.0-1.9 cm long. **Sepals** similar, bases attenuate, margins slightly sinuate, apices acute, membranaceous, yellow, densely spotted with brown, dorsal sepal obovate, 6.0-7.0 x 2.0-3.0 mm; lateral sepals (1\3) adnate, elliptic, 8.0-9.0 x 2.0-3.0 mm. **Petals** obovate, base briefly angled, margins entire, apex emarginate, membranaceous, completely yellow or spotted with brown only at the base, 1.0-1.4 x 1.0-1.2 cm. **Lip** yellow with central region of callus with light brown spots, 1.2-1.6 x 1.2-1.6 cm; lateral lobes elliptical to oval, yellow, 2.0-4.0 x 2.0-5.0 mm; median lobe ovate, margin slightly erose, straight to emarginate, yellow, 0.70.9 x 1.2-1.6 cm; isthmus 2.0-4.0 cm; disc region with complex callus involving several verrucous protuberances and lamellae. **Column** 3.0 x 1.0-2.0 mm, brownish, infra-stigmatic tubule 1.0-2.0 x 1.0-2.0 mm, wings or yellow stellidia 1.0 x 1.0 mm. **Fruit** not seen.

Alatiglossum macropetalum (Lindl.) Baptista occurs in Bolivia and Brazil (Tropicos, 2014). In Brazil, it is distributed in the Federal District and in the states of Goiás, Maranhão, Mato Grosso, Mato Grosso do Sul, Minas Gerais, Parà, São Paulo and Tocantins, in the Central-West, Northeast, North and Southeast regions, in the Cerrado and Atlantic Forest biomes. It grows in Ciliary Forest or Gallery Forest, Dry Semideciduous Forest, Ombrophilous Forest (= Rain Forest) (Barros *et al.,* 2015). In the Federal District, it grows in riparian forest and semi-deciduous dry forest, flowering in July and August.

Selected material: **APA de Cafuringa,** Near the community of Córrego do Ouro, Vni.2003, *Pellizzaro, K.F. et al.,* 27 (CEN). **FERCAL,** 6.7 km from the CIPLAN cement factory, VII.1990, *Bianchetti, L. 869* (CEN). **Lajinha,** Estrada de Sobradinho, Vii.1965, *Sucre, D. 647* (UB).

Additional material examined: MINAS GERAIS. **Unai,** about 1 km downstream of the spillway, Vii.2003, *Santos, A.A. 2056* (CEN).

2. *Aspasia* Lindl., *Gen. Spec. Orch. Pl.,* 139 (1833).

Epiphytic **herbs.** Short **rhizome** between secondary stems. Secondary stems swollen into **pseudobulbs**, oval, ancipitate, partially covered with bracts, topped by 1-2 leaves. **Leaves** conduplicate, bifacial, leathery, articulated, green. Lateral, raceme-like **inflorescence,** shorter than the length of the leaves, 1-6 flowers. **Flowers are** resupinate, showy, cream-colored with pink or darker brown spots. **Sepals** and **petals** free and similar in shape and size. **Lip** trilobed, panduriform, ecalcarate, inserted in the median part of the column, forming a cavity (pseudo nectary), lateral lobes as wide as the median one; region of the disc with a callus composed of two thickened parallel veins. **Column** not winged, straight; absence of infrasigmatic tubule; clinandrium poorly developed, smooth. **Fruit** elliptic to ovoid capsule, not slightly 3-angled; mature fruit with apex of carpels united.

A genus established by the English botanist John Lindley in 1833, who in *Obra princeps* did not describe the meaning of the name *Aspasia,* but in Chase's work (2009) reference is made to Aspasia, the beautiful and cheerful wife of Pericles.

The genus has seven species distributed in Guatemala, Belize, Panama, Venezuela, Colombia, Paraguay, Ecuador, Peru and Brazil (Chase, 2009). In Brazil, five species occur in the Amazon, Cerrado and Atlantic Forest biomes. The Cerrado biome is represented by just one species, *Aspasia variegata* Lindi., which is present in the Federal District (Barros *et al.,* 2015).

Aspasia species have a flat lip with the base attached to the column at a right angle, forming a false nectary. They are pollinated by male Euglossini bees, probably using deception strategies (absence of nectar) or by fragrance-collecting bees, depending on the species (Zimmerman & Aide, 1987).

2.1. *Aspasia variegata* **Lindl.** Edwards's Botanical Register 22: t. 1907. (1836). **Fig. 1.** G-H; **4.** B.

Light green **pseudobulbs**, 3.3-7.7 x 1.4-1.8 cm, protected by 2-3 persistent, articulated sheaths, 11.5-19.1 cm long and paleaceous sheaths, sessile **leaves,** narrowly elliptic to oblanceolate, base attenuated, apex acute, 10.2-24.0 x 1.3-3.5 cm. **Inflorescence** 2-4 flowers, 5.0-9.5 cm long overall. **Flowers** pedicellate, pedicels 0.7-1.1 cm long. **Sepals** similar, elliptic-obovate, base attenuated, margins entire, apex acute-apiculate, membranous, yellow with violet stripes, 1.9-2.0 x 0.4-0.7 cm. **Petals** more obovate than sepals, base attenuated, margins entire, apex acute, membranous, yellow with violet stripes, 1.7-2.0 x 0.7-0.8 cm. **Lip** yellow with violet stripes, 1.5-1.6 x 1.8-2.0 cm; lateral lobes oval, yellow with violet stripes, 0.5-0.6 x 0.9-1.0 cm; median lobe oval-subquadrangular, sinuate margin, emarginate apex, 0.9-1.0 x 1.2-1.6 cm; region of biconical disc, linear callus. **Column** 1.4-1.6 cm, brown. **Fruit** not seen.

Aspasia variegata occurs in South America, including Bolivia, Colombia, French Guiana, Suriname, Trinidad, Venezuela and Brazil (Williams, 1974). In Brazil it is distributed in the Federal District and in the states of Amapà, Amazonas, Goiàs, Maranhao, Mato Grosso, Parà, Rondônia, Roraima and Tocantins, in the Central-West, Northeast and North regions, and in the Amazon and Cerrado biomes. It grows in Campinarana, Ciliary Forest or Gallery Forest, Igapó Forest, Terra Firme Forest, Vàrzea Forest, Deciduous Dry Forest, Semideciduous Dry Forest (Barros *et al.,* 2015). In the Federal District, it grows in mesophytic forest (Dry Deciduous Forest) with flowering in October.

Material examined: **DF-205,** Fazenda Acre, Fercal region, X.1999, *Augusto, M.M.* 1 (CEN).

Additional material examined: MINAS GERAIS. **Unai,** Forest next to the escape tunnel, II.2002, *Santos, A.A. 1618* (CEN).

3. *Cohniella* Pfitzer, *Nat. Pflanzenfam.* 2, 6: 194. 1889.

Epiphytic or rupicolous **herbs.** Short **rhizome** between secondary stems. Secondary stems swollen into **pseudobulbs,** small (in relation to the length of the leaf), suborbicular, protected by scarlet sheaths (thin, dry and membranous), topped by a leaf. **Leaves** terete, unifacial, fleshy, articulated, often green with purple or red dots. **Inflorescence** lateral, usually paniculate, rarely racemose, longer than the leaves, few to multiflorous. Showy resupinate **flowers,** usually yellow and white with brown spots. **Sepals** and **petals** free and similar, usually green or yellow with red or brown dots or spots. **Lip** trilobed, echolobed, base of lip inserted perpendicularly to the base of the column, apical lobe usually much larger than the lateral ones; disc region with callus composed of series of plates and/or teeth. **Column** relatively short, with semi-circular or stellate wings flanking the stigma; presence of a prominent infra-stigmatic tubule at the base; clinandrium poorly developed, smooth. **Fruit** ellipsoid to suborbicular capsule, not to slightly 3-angled; mature fruit with apex of carpels united.

Genus established by the German botanist Ernest Hugo H. Pfitzer, in 1889, in allusion to the similarity with the species *Cohnia quekettiodes* Rcchb.f. (1852) (Fernândez-Concha, 2010).

The genus has around 17 species (known as rat-tail oncidiums, due to the morphology of the leaves), widely distributed from northern Mexico to northern Argentina and Brazil (Cetzal Ix *et al.,* 2012). In Brazil, five species occur in almost all biomes, except the Pampa biome (Cetzal Ix *et al.,* 2012). Two species are represented in the Cerrado biome, *Cohniella cepula* (Hoffmanns.) Carnevali & G. Romero and *Cohniella jonesiana* (Rchb.f.) Christenson, which also occur in the Federal District (Barros *et al.,* 2015).

Cohniella species usually have yellow and some white flowers with brown spots and offer oils and resins as rewards and are visited by bees of the genera *Centris* and *Paratetrapedia* (Silvera, 2002 - unpublished - *apud* Neubig *et al.,* 2012) and pollinated by bees of the genus *Trigona* (Parra-Tabla *et al.,* 2000). Some species do not offer any kind of reward and are probably pollinated by mistake.

KEY TO *COHNIELLA* SPECIES FOR THE FEDERAL DISTRICT

2. Leaves generally erect; sepals and petals generally shorter than the lip (petals and sepals = 0.6-1.5 cm; lip = 1.3-1.6 cm); lip yellow; lip callus tridentate ____________ 1. *C. cepula*

1. Leaves always pendulous; sepals and petals usually the same length or slightly shorter than the

lip (petals and sepals = 1.6-2.5 cm; lip = 2.0-2.2 cm); lip white; lip callus pentadentate 2.   *C. jonesiana*

1.1. ***Cohniella cepula*** **(Hoffmanns.) Carnevali & G.A. Romero.** Brittonia 62(2): 167 (2010).

Fig. 1. A-B; **4.** B.

Green **pseudobulbs**, 0.5-0.7 x 0.3-0.7 cm, protected by persistent, small palea sheaths. **Leaves** sessile, narrowly linear, truncate bases, acute apices, succulent, green, 2.5-17.5 x 0.2-1.0 cm. Paniculate **inflorescence**, 10-25 flowers, 32.0-72.5 cm long overall. **Flowers** pedicellate, pedicels, 0.9-1.6 cm long. **Sepals** similar, rhomboid to broad-elliptic, bases attenuated, margins entire, apices obtuse, membranaceous, yellowish densely spotted with reddish brown, dorsal sepal, 0.6-1.0 x 0.3-0.7 cm; lateral sepals free, 0.8-1.5 x 0.4-0.7 cm. **Petals** obovate, bases attenuated, margins slightly wavy, apices obtuse, membranaceous, yellowish, densely spotted with reddish brown, 0.6-0.9 x 0.3-0.6 cm. **Lip** trilobed, yellow, 1.3-1.6 x 1.1-1.7 cm; lateral lobes obovate to triangular, in the same plane as the median lobe, yellow, 2.0-5.0 x 4.0-7.0 mm; median lobe obovate to spatulate, margin slightly erose, apex emarginate, 0.6-0.9 x 1.1-1.7 cm; isthmus yellow, 3.0-5.0 mm; disc region with tridentate callus and two lateral keels. **Column** 2.0-4.0 x 1.0 mm, brown, infra-stigmatic tubule 1.0-2.0 x 1.0 mm, wings or stellidia 1.0-2.0 x 1.0-2.0 mm. **Fruit** not seen.

Cohniella cepula occurs in Argentina, Bolivia, Paraguay, Peru and Brazil (Cetzal Ix *et al.*, 2012). In Brazil, it is distributed in the Federal District, and in the states of Acre, Goiás, Mato Grosso, Mato Grosso do Sul, Rondônia and Tocantins, in the Central-West, North and Northeast regions and in the Amazon, Caatinga and Cerrado biomes. It grows in Mesophytic forests (Barros *et al.*, 2015). In the Federal District, it grows in Ciliary Forest and Dry Deciduous Forest, and blooms especially between April and June.

Selected material: **APA de Cafuringa,** Near Repùblica dos Urubus, V.2006, *Pellizzaro, K.F. 30* (CEN). **FERCAL,** Limestone outcrop on the left DF-205, XII.2000, *Batista, J.A.N. 1101* (CEN). **Nùcleo Rural Sobradinho II,** Chacarà Tupa-Ipê, IV.2001, *Mendes, R.A. 339* (CEN).

1.2. ***Cohniella jonesiana*** **(Rchb. f.) Christenson.** Lindleyana 14(4): 177. 1999.

Green **pseudobulbs**, 0.6-1.1 x 0.3-1.0 cm, protected by persistent, small palea sheaths. **Leaves** sessile, narrowly linear, truncate bases, acute apices, succulent, dark green, 2.2-40.0 x 0.2-2.0 cm. Paniculate **inflorescence**, 2-10 flowers, 30.0-70.0 cm long overall. **Flowers** pedicellate, pedicels, 2.0-2.5 cm long. **Sepals** similar, obovate to oblanceolate, bases rather attenuated, margins entire, crenate, apices obtuse, membranaceous, white densely spotted with reddish-brown; **dorsal sepal,** 1.6-2.5 x 0.8-1.3 cm; **lateral sepals** partially fused at the base, 1.8-2.4 x 0.8-1.2 cm. **Petals** oblong to oblanceolate, bases attenuated, margins slightly wavy, apices obtuse, membranaceous, white densely spotted with reddish brown, 2.0-2.6 x 0.8-1.0 cm. **Lip** white spotted with red at the base, 2.0-2.5 x 2.0-2.6 cm; lateral lobes oblong, in the same plane as the median lobe, white, margins erose, 0.3-0.6 x 0.5-0.9 cm; median lobe obovate to spatulate, margin erose to lacerate, apex emarginate, 1.2-2.0 x 2.0-2.6 cm; isthmus yellow, 1.0-3.0 cm; disc region with pentadentate callus, central tooth compressed laterally flanked by four smaller teeth (two posterior and two anterior). **Column** 6.0 x 2.0-4.0 mm, brown, presence of infra-stigmatic tubule 2.0-3.0 x 1.0-3.0 mm, wings or stellidia 1.5-2.0 x 1.0-3.0 mm. **Fruit** not seen.

Cohniella jonesiana occurs in Argentina, Bolivia, Brazil and Paraguay (Cetzal Ix *et al.,* 2012). In Brazil, it is distributed in the Federal District, and in the states of Mato Grosso, Minas Gerais, Paranà and Sao Paulo, occurring mainly in the Midwest, Southeast and South regions, and in the Cerrado biome (Cetzal Ix *et al.,* 2012). It grows in riparian or gallery forests and mesophytic forests (Barros *et al.,* 2015). According to Pellizzaro *et al.* (2004), this species grows in the Federal District in dry deciduous forest and flowers from December to February.

OBS: No herbarium *vouchers* were found for this species. However, works such as "Estudo do gènero *Oncidium"* (Pellizzaro *et al.,* 2004) and "A Familia Orchidaceae na APA de Cafuringa e sua contribuiçao para a diversidade da flora do Distrito Federal e do bioma Cerrado" (Bianchetti *et al.,* 2005) record the occurrence of the species in the Federal District.

The botanical description was based on additional material examined from Minas Gerais, with adaptations of the description contained in the work by Cetzal Ix *et al.* (2012).

Additional material examined: MINAS GERAIS. **Unai,** Gruta do Tamboril region, X.2007. *Miranda, Z.J.G. 93* (CEN).

4. *Comparettia* Poepp. & Endl., *Nov. Gen. Spec.* 1, 42, t. 73 (1835).

Epiphytic **herbs.** Short **rhizome** between secondary stems. Secondary stems swollen into **pseudobulbs**, elongated to orbicular, topped by 1-4 leaves. **Leaves** conduplicate, bifacial, leathery, articulated, usually green, sometimes pigmented dark red. Lateral, raceme-like **inflorescence,** usually exceeding the length of the leaves, 5-25 flowers. Red, orange, yellow to pink resupinate, showy **flowers.** Dorsal **sepal** free, the lateral ones fused to form a nectariferous spur. **Petals** free. **Lip** hastened, entire, calcarate, base of the lip inserted perpendicularly to the base of the column, base with two divergent filiform prolongations (positioned inside the spur formed by the lateral sepals), apical lobe flattened, well developed; disc region with a simple, raised callus. **Column** not winged, straight; absence of infrasigmatic tubule at base; clinandrium poorly developed, smooth. **Fruit** elliptic to ovoid capsule, not to slightly 3-angled; mature fruit with apex of carpels united.

A genus established by the German botanist Eduard Friedrich Poeppig and the Hungarian botanist

Stephan Ladislaus Endlicher, in 1835, in honor of Andrea Comparetti, a physiologist and professor of botany in Padua, Italy (Chase, 2009).

The genus *Comparettia* has around 60 species distributed from Mexico and the West Indies to South America, encompassing Bolivia, Peru and Brazil (Chase, 2009). In Brazil, five species occur in the Amazon, Cerrado and Atlantic Forest biomes. Only one species occurs in the Cerrado biome, *Comparettia coccinea* Lindl., which is also present in the Federal District (Barros *et al.*, 2015).

Comparettia species have flowers with a spore-shaped nectary formed by the lateral sepals. This spur is supplied by a pair of glands at the base of the column which secrete nectar. Hummingbird pollination has been documented for *C. falcata*. Pollination by butterflies and long-tongued bees seems likely for some species (Dodson, 1965; Salguero-Faria & Ackerman, 1999). Pansarin *et al.* (2012) presented preliminary results showing that *Comparettia coccinea* has red flowers with a long recurved spur formed by the fusion of the lateral sepals. The inside of the spur has unicellular trichomes responsible for nectar secretion. This species is pollinated by the butterflies *Heliconius ethilla narcaea* (Godart, 1819) and *H. erato phyllis* (Fabricius, 1775).

4.1. *Comparettia coccinea* **Lindl.** Sketch Veg. Swan 14: t. 68. (1838).

Fig. 1. C-D; **4.** C.

Oblong, green **pseudobulbs**, 0.8-2.5 x 0.2-0.5 cm, topped by 1 leaf; protected by persistent palea sheaths. **Leaves** sessile, lanceolate, bases

attenuated, obtuse apices, green, 8.2-21.2 x 0.8-1.5 cm. **Inflorescence** 1-6 flowers, total length 8.331.0 cm. **Flowers** pedicellate, pedicels, 1.1-1.5 cm long. **Sepals** similar, ovate, base slightly attenuated, margin entire, apices acute to mucronate, membranaceous, orange-red, dorsal sepal 8.0 x 4.0 mm, lateral sepals fused, 5.0-6.0 x 3.0 mm, spur 2.0 cm. **Petals** oval, bases attenuated, margins entire, apices acute, membranous, orange-red, 5.0-7.0 x 3.0-4.0 mm. **Lip** obovate, cuneate base, sinuate margin, emarginate apex, orange-red with slightly yellowish underside, 1.6 x 1.6 cm; disc region with bilamellar callus. **Column** 2.0-3.0 mm, whitish. **Fruit** 2.0 x 1.5 cm, ellipsoid.

Comparettia coccinea occurs in Bolivia, Peru and Brazil (Tropicos, 2014). In Brazil it is distributed in the Federal District and the states of Bahia, Espirito Santo, Minas Gerais, Pernambuco, Rio de Janeiro and Sao Paulo, in the Central-West, Northeast and Southeast regions and in the Cerrado and Atlantic Forest biomes. It grows in Ciliary Forest or Gallery Forest, Dry Semideciduous Forest and Ombrophilous Forest (= Rain Forest) (Barros *et al.*, 2015). In the Federal District, it grows in riparian forest, non-flooded gallery forest and flooded gallery forest, with flowering from January to March and fruiting from March onwards.

Selected material: **Brasilia National Park,** Mata dos Três Barras, Ш.1999, *Santos, A.A. 398*

(CEN). **Guarà Ecological Reserve,** 15°50'00"S, 47°57'00"W, X.1994, *Oliveira, R.S. 16* (UB). **Taquatinga,** Floresta Nacional de Brasilia, 15°79'02"S, 48°08'64"W, X.2012, *Lima, J.H. 16* (UB). **Vàrzea Bonita,** II.1978, *Heringer, E.P. 16751* (IBGE).

5. *Coppensia* Dumort, *Nouv. Mém. Acad. Sci. Bruxelles*, 9, 10 (1835).

Terrestrial, epiphytic or rupicolous **herbs.** Short to elongated **rhizome** between secondary stems. Secondary stems swollen into **pseudobulbs,** fusiform to oval, usually rough and ancipitate, partially covered by leaf sheaths, topped by 2-4 leaves. **Leaves are** conduplicate, bifacial, leathery, articulated, green. Lateral **inflorescence,** raceme or panicle type, longer than the leaves, pauci to multiflora. **Flowers are** resupinate, spreading and showy, usually yellow with brown or red spots. Dorsal sepal free; lateral sepals fused in various sizes. Free **petals,** similar to the sepals in size, shape and color. **Lip** trilobed, echolobed, base of lip inserted perpendicularly to base of column, lateral lobes auriculate, middle lobe usually quite broad; disc region with multi-verrucose or lamellate callus, margins of disc and isthmus entire, never jagged or fimbriated. **Column** with semi-circular or stellate wings flanking the stigma; presence of a prominent infra-stigmatic tubule at the base; clinandrium poorly developed, smooth. **Fruit** elliptic to ovoid capsule, not to slightly 3-angled; mature **fruit** with apex of carpels united.

A genus established by the Belgian botanist Barthéleny C.J. Dumortier in 1835, but there is no reference to its etymology in the *Obra princeps.* As the surname Coppens is a common surname in Belgium and France, it is possibly a tribute to Aurèle Augustin Coppens (1668 - 1740), a Belgian painter and artist who created a series of famous tapestries illustrating the destruction of Brussels by French troops in 1695.

The genus has around 48 species (WCSP, 2014) with a Neotropical distribution from Bolivia, Paraguay, Argentina, Uruguay and Brazil. In Brazil there are 38 species (29 endemic) in all biomes, but with very low representation in the Amazon and Caatinga biomes. In the Cerrado biome, 15 species are represented and of these, two occur in the Federal District (*Coppensia hydrophila* (Barb.Rodr.) Campacci and *Coppensia varicosa* (Lindl.) Campacci (Barros *et al.,* 2015).

Much of the literature deals with species of *Gomesa s.l.* (today including species of *Coppensia* according to Docha Neto, 2007). Therefore, many observations, especially regarding the reproductive biology of the species, will be referred to as belonging to the genus *Gomesa.*

Species of the genus *Gomesa* have great floral diversity and reproductive biology (Dressler, 1993). Molecular studies (Williams *et al.,* 2001; Chase *et al., 2009)* have shown that the "Gomesa clade" is monophyletic and that the characteristic of offering oils as a reward to pollinators is well represented (Singer *et al.,* 2006). The production of oils is related to specialized glands called elaiophores. Aliscioni *et al.* (2009) and Torreta *et al.* (2011) demonstrated that *Gomesa bifolia* (Sims) M.W.Chase & N.H.Williams (=*Coppensia bifolia*) has elaiophores in the callus and that the structure of the secretory cells is similar to those described for other Oncidiinae species. Torreta *et al.* (2011) showed that *Gomesa bifolia* has no fragrance and is a self-incompatible, non-autogamous species, dependent on pollinators and that female *Centris trigonoides* bees (Apidae, Centridini) were the exclusive pollinators. Contrary to the other generalist comments, Pansarin *et al.* (2012) presented preliminary results showing that *Gomesa varicosa* does not have any type of floral resource-producing gland (also confirmed by Stpiczynska *et al.*, 2013) and is pollinated by mistake by *Centridini* bees (Epicharis (Epicharana) *flava* (Friese, 1900) and *Centris* sp.).

Coppensia species are pollinated by bees and generally occur where there is a high concentration of these orchids (Miller *et al.*, 2006).

KEY TO *COPPENSIA* SPECIES FOR THE FEDERAL DISTRICT

1. terrestrial herbs, raceme inflorescence 1. *C. hydrophila*

1. epiphytic herbs, panicle-like inflorescence 2. *C. varicosa*

5.1. *Coppensia hydrophila* **(Barb. Rodr.) Campacci.** Bol. CAOB 62: 55 (2006). **Fig. 1.** E-F; **4.** C.

Terrestrial **herb.** Greenish **pseudobulbs**, 1.9-5.0 x 0.9-2.6 cm, laterally protected by persistent, articulated leaf sheaths, 8.0-42.5 cm long, topped by 2-3 leaves. **Leaves** sessile, narrowly elliptic to oblanceolate, bases attenuated, apices acute, 10.2-24 x 1.3-3.5 cm. **Inflorescence** raceme, 5-40 or more flowers, 0.6-1.0 m long overall. **Flowers** pedicellate, pedicels, 0.7-2.0 cm long. **Sepals** similar, elliptic, bases slightly narrowed, margins entire, apices obtuse, membranaceous, completely yellow or lightly spotted with vinaceous brown, dorsal sepal 5.0-7.0 x 2.0-4.0 mm; lateral sepals 1\3 adnate, 5.0-9.0 x 1.03.0 mm. **Petals** obovate, bases slightly narrowed, margins entire, apices acute, apiculate, membranaceous, entirely yellow or lightly spotted with vinaceous brown, 5.0-7.0 x 3.0-5.0 mm. **Lip** yellow, 1.4-1.8 x 1.6-2.0 cm; lateral lobes transversely elliptic, yellow, 1.0-4.0 x 2.0-4.0 mm; median lobe oval, margin sinuate, apex emarginate, 0.8-1.3 x 1.6-2.0 cm; isthmus 2.0-3.0 mm, yellow; **disc region** with verrucous callus involving dentiform structures and ridges (median region bidentate; anterior region tri-crested). **Column** 3.0-5.0 x 1.0-1.5 mm, brown, infra-stigmatic tubule 2.0 x 1.0-2.0 mm, wings or stellidia 2.0-3.0 x 1.0-2.0 mm. **Fruit** not seen.

Coppensia hydrophila occurs in Paraguay and Brazil (Tropicos, 2014). In Brazil, it is distributed in the Federal District and in the states of Bahia, Espirito Santo, Goiás, Minas Gerais, Paranà, Pernambuco, Rio de Janeiro, Rio Grande do Sul, Santa Catarina, São Paulo and Tocantins, in the Central-West, Northeast, Southeast and South regions, especially in the Cerrado biome. It grows in environments rich in moisture, such as Campos de Altitude, Campo Limpo and Campo Rupestre

(Barros *et al.*, 2015). In the Federal District, it grows in wetland, swamp and rocky field environments, and blooms from October to January.

Selected material: **Below Velhacap,** Close to Nucleo Bandeirante, XI.1978, *Ratter, J. 4293* (UB). **Fazenda Agua Limpa da UnB,** Córrego Taquara, 15°36"S, 47°54"W, XI.2003, *Pinagé, E.R. s.n.* (UB14529). **Fundaçâo Zoobotànica de Brasilia,** XI.1961, Heringer, E.P. 8447 (NY). **Lago Sul,** Campo as margens do Córrego do Gama, X.1998, *Batista, J.A.N. 800* (CEN). **Boca da Mata Park,** XI.1995, *Rezende, J.M. 255* (CEN). **Guarà Ecological Reserve,** X.1990, *Batista, J.A.N. 118* (CEN). **IBGE Ecological Reserve,** X.1979, *Heringer, E.P. 3744* (IBGE). **Riacho Fundo Ecological Sanctuary,** XI.1994, *Oliveira, R.S. 28* (UB). **Industrial Sector,** XII.1965, *Belém, R.P. 1896* (UB).

Oncidium hydrophilum var. *immaculatum* L.C. Menezes, is also cited for the Federal District and considered, by the List of Species of the flora of Brazil (2014), as a synonym of *Coppensia hydrophila* (Barb. Rodr.) Campacci. An examination of the type material (UB! Brasilia. Reserva Ecològica do Guarà. X.1990, *Menezes, L.C. 8*) showed that there were no flowers or inflorescences.

Still on the subject of var. *immaculatum*, several authors consider that the absence of anthocyanins in two plants in a population would be better treated as a form and not as a variety (Christenson, 1996; Barros & Batista, 2004).

5.2. *Coppensia varicosa* (Lindl.) Campacci Bol. CAOB 62: 57 (2006).

Fig. 4. C.

Epiphytic **herb. Pseudobulbs** ovate-oblong, green with burgundy-painted furrows, 3.5-7.8 x 0.9-2.8 cm, protected by articulated, persistent leaf sheaths, 7.1-15.5 cm long, topped by 2-3 leaves. **Leaves** elliptic to obovate, base attenuated, apex acute, 16.0-30.0 x 1.0-4.0 cm. Panicle-like **inflorescence**, 20 to just over 100 flowers, 80.0-116.0 cm long overall. **Flowers** pedicellate, pedicels 1.2-3.1 cm long. **Sepals** similar, elliptic, bases slightly narrowed, margins entire, apices acute, membranaceous, yellow spotted with reddish brown, dorsal sepal 0.7-1.2 x 0.3-0.6 cm; lateral sepals flattened 1/3 to half, 0.8-1.5 x 0.20.6 cm. **Petals** oval, bases attenuated, margins slightly sinuous, apices obtuse, mucronate, membranaceous, yellow spotted with reddish brown, 0.8-1.2 x 0.3-0.7 cm. **Lip** yellow, 2.3-3.5 x 2.2-4.1 cm; lateral lobes obovate to subrotund, yellow, 0.3-0.5 x 0.3-0.5 cm; median lobe broad, reniform, slightly constricted towards the apex giving the impression of being tetra-lobed, apex emarginate, 1.4-3.0 x 2.2-4.1 cm; disc region usually painted reddish-brown with verrucous callus involving dentiform structures and ridges (median region bidentate; anterior region tri-crested). **Column** 4.0-5.0 x 1.0 mm, brown, infra-stigmatic tubule 2.0 x 1.0 mm, wings or stellidia 3.0 x 1.0-2.0 mm. **Fruit** not seen.

Coppensia varicosa occurs in Bolivia, Paraguay and Brazil (Tropicos, 2014). In Brazil it is distributed in the Federal District and the states of Bahia, Goiàs, Minas Gerais, Paranà, Santa Catarina and Sao Paulo, in the Central-West, Northeast and Southeast and South regions and in the Caatinga, Cerrado and Atlantic Forest biomes. It grows in the Rupestrian Field, Riparian Forest or Gallery Forest, Dry Semideciduous Forest, Ombrophilous Forest (= Rain Forest) (Barros *et al.,* 2015). In the Federal District, it grows in riparian and gallery forests and flowers mainly from January to April.

Selected material: **ARIE Cerradao,** 15°51"S, 47°49"W, I.2008, *Silva, J.S. 318* (UB). **São Bartolomeu River Basin,** VII.1979, *Heringer, E.P. 1831* (IBGE). **Cabeça de Veado,** III.1961, *Heringer, E.P. 8065* (UB). **Catetinho,** II.1979, *Salles, A.E.H. 3* (IBGE). **EEJBB, JBB water catchment,** 15°52S, 47°51"W, IV.2003, *Soares, A.C.A. s.n.* (HEPH29268). **Fazenda da Universidade de Brasilia (Fazenda**

Agua Limpa), Mata do Córrego do Gama, III.1992, *Bianchetti, L. 1239* (CEN). **Fazenda Sucupira,** Córrego do Açudinho, 15°55'00"S, 48°01'00"W, II.2000, *Faria, J.G. 254* (CEN). **Grota behind Gama cemetery,** 16°01' S, 48°03'W, IX.1996, *Gomes, S.M. 86* (CEN). **IBGE Ecological Reserve,** Taquara Stream. 15°56'S, 47°55'W, III.1994, *Walter, B.M.T. 2091* (CEN).

6. *Ionopsis* Kunth, *Nov. Gen. Spec. Pl.*, 1, 348, t. 83 (1816).

Epiphytic **herbs.** Short **rhizome** between secondary stems. Secondary stems swollen into **pseudobulbs,** inconspicuous, always covered by leaf sheaths, topped by 1 leaf. **Leaves are** conduplicate, bifacial, leathery, articulated, green or purplish. Lateral panicle-like **inflorescence,** longer than the leaves, single to multiflora. **Flowers are** resupinate, spreading and showy, green, pink to violet. Dorsal sepal free; lateral sepals partially fused and at the base forming a small mentum. Free **petals,** similar to the sepals. **Lip** entire, obovate, echolocate, base of lip inserted perpendicularly to base of column, mid-apical region very dilated, apex emarginate; disc region with bilamellar callus. **Column** not winged, slightly dilated at the apex; absence of infrasigmatic tubule; clinandrium poorly developed, smooth. **Fruit** elliptic to ovoid capsule, not to slightly 3-angled; mature fruit with apex of carpels united.

A genus established by the German botanist Carl Sigismund Kunth in 1816, in allusion to the superficial similarity between the flowers of the first species described for the genus and those of violets (Greek: *ion* = violet and *opis* = appearance, similarity) (Chase, 2009).

The genus has three species, widely distributed in the Neotropics, from southern Florida to Central and South America (Chase, 2009). In Brazil, three species occur in the Amazon, Cerrado and Atlantic Forest biomes. In the Cerrado biome, and in the Federal District, it is represented by just one species, *Ionopsis Utricularioides* (Benth.) Lindl. (Barros *et al.*, 2015).

Ionopsis species have pink or white flowers with an entire lip, well developed in relation to the other floral parts, with a small mentum formed by the lateral sepals. The flowers offer no reward and are probably pollinated by mistake by bees looking for nectar (Neubig *et al.*, 2012). Work carried out by Aguiar (2014) shows that *I. utricularioides* has osmophores in the protuberances located at the base of the lip and, around them, and secretory papillae, capable of producing certain fragrances that attract bees, but it has not yet been recorded which types of bees.

6.1. *Ionopsis utricularioides* (Sw.) Lindl. Coll. Bot. 8: t. 39, f. A. (1826).

Fig. 3. E-F; **4.** D

Pseudobulbs elliptical to ovoid, green, 0.4-2.7 x 0.1-0.8 cm, hidden by sheaths. **Leaves** or leaf sheaths with developed limb, persistent, oblong or linear- ligulate 5.0-15.2 x 0.8-1.2 cm. **Inflorescence** 9-150 flowers or more, 21.0-66.0 cm long overall. **Flowers** pedicellate, pedicels, 0.4-1.0 cm long. **Sepals** similar, ovate to elliptic, bases attenuate, apices acute, margins entire, membranaceous, lilac, dorsal sepal 4.0-5.0 x 1.0-2.0 mm; lateral sepals fused halfway, 4.05.0 x 1.0-2.0 mm. **Petals** obovate, bases attenuated, apices obtuse, margins entire, membranous, lilac, 4.0-6.0 x 2.0-3.0 mm. **Lip** lilac to white, cuneate base, entire margin, emarginate apex, 0.8-2.0 x 1.0-1.5 cm. **Column** 1.0-2.0 mm, brownish. **Fruit** not seen.

Ionopsis utricularioides occurs in parts of North, Central and South America, from Florida (USA) to Argentina and Brazil (Aguiar, 2014). In Brazil it is distributed in the Federal District and the states of Acre, Amazonas, Bahia, Goiàs, Parà, Paranà, Pernambuco, Maranhao, Mato Grosso, Mato Grosso do Sul, Minas Gerais, Rio de Janeiro, Rondônia, Sergipe, Santa Catarina and Sao Paulo, in the Central-West, Northeast, North, Southeast and South regions, and in the Amazon, Cerrado and Atlantic Forest biomes, in natural conditions or in anthropized environments. It grows in Ciliary Forest or Gallery Forest, Igapó Forest, Terra Firme Forest, Deciduous Dry Forest, Perennial Seasonal Forest, Semideciduous Dry Forest, Ombrophilous Forest (= Rain Forest) (Barros *et al.*, 2015). In the Federal District, it occurs in flooded gallery forest and mesophytic forest (Dry Deciduous Forest), and flowers from June to December.

Selected material: **APA de Cafuringa,** Repûblica dos Urubus, XII.2002, *Pellizzaro, K.F. 28* (CEN). **FERCAL,** APA de Cafuringa, VIII.1990, *Vieira, R. 361* (CEN). **Jardim Botànico de Brasilia,** 15°52'S, 47°51' W, VIII.2000, *Nóbrega, M.G. 1294* (HEPH). **Guarà Ecological Reserve,** VIII.1999, *Batista, J.A.N. 935* (CEN). Margin of the **Ribeirao Jacaré,** cut by the DF-100, VI.1991, *Bianchetti, L. sn.* (CEN15096).

7. *Lockhartia* Hook, *Bol. Mag.*, 54, t. 2715 (1827).

Epiphytic **herbs.** Short **rhizome** between secondary stems. Secondary stems not swollen into **pseudobulbs**, erect or pendulous, covered in short leaves. **Leaves** unifacial, fleshy, bases equitant or imbricated, **non-articulate**, green. **Inflorescence** short, lateral or terminal, raceme or panicle type, length less than or equal to the leaves, pauciflorous; bracts enlarged, semi-orbicular. **Flowers** usually resupinate, small, usually yellow with red or dark brown spots. Free **sepals** and **petals**, similar in size and shape. **Lip** usually trilobed, echolobed, base of lip inserted perpendicular to base of column, apical lobe bilobed, oblong to almost circular; lateral lobes often linear and arched; disc region with simple to complex callus, tuberculate or horned. **Column** usually winged or with small wings flanking the stigma; absence of infra-stigmatic tubule; clinandrium poorly developed, smooth. **Fruit** ellipsoid to suborbicular capsule, not 3-angled; mature fruit with free and reflexed carpel apex.

A genus established by the English botanist William Jackson Hooker in 1827 in homage to David Lockhart, superintendent of the Imperial Botanical Gardens of Trinidad and Tobago during the 18th century (Chase, 2009).

The genus has around 30 species with a Neotropical distribution, extending from North America (Mexico) to South America (Venezuela, French Guiana, Bolivia, Peru, Trinidad and Brazil) (Chase, 2009). In Brazil there are seven species (four endemic), distributed in the Amazon, Atlantic Forest and Cerrado biomes. Only one species occurs in the Cerrado biome, *Lockhartia goyazensis* Rchb.f., which is present in the Federal District (Barros *et al.,* 2015).

There is no information on pollination mechanisms or pollinators for *Lockhartia* species. However, the floral morphology, especially the callus morphology, indicates that they belong to the Oncidiinae group with flowers that produce oils and are probably pollinated by oil-collecting bees, similar to other *Oncidium* species that exhibit this syndrome (Chase, 2009).

7.1. *Lockhartia goyazensis* Rchb.f. Bot. Zeitung (Berlin) 10: 768 (1852).

Fig. 2. A-B; 4. D.

Leaves sessile, triangular in lateral view and broadly ovate in distended frontal view, truncate base, acute to slightly rounded apex, 1.0-2.3 x 0.3-1.4 cm. **Inflorescence** raceme, 2-4 flowers, 1.0-1.6 cm long overall; 2-4 petaloid bracts, patent, oval, acute apex, light brown, 2.0-4.0 mm long. **Flowers** pedicellate, pedicels 1.0-7.0 mm long. Dorsal and lateral **sepals** obovate, bases truncate, margins entire, apices obtuse, membranaceous, yellow, 3.0-4.0 x 2.0-3.0 mm; **petals** elliptic, bases truncate, apices rounded, margins entire, membranaceous, yellow, 3.0-4.0 x 2.0-3.0 mm. **Lip** yellow with brown spots in the center, 5.06.0 x 7.0-9.0 mm; lateral lobes linear, yellow, 3.0-4.0 x 0.8 mm; median lobe obovate-subquare, margin straight, apex emarginate, 3.0-4.0 x 3.0-5.0 cm; disc region with complex, tuberculous callus, occupying the basal and central part of the median lobe. **Column** 2.0 x 1.0 mm, brown. **Fruit** 11.0 x 6.0 mm, oblong.

Lockhartia goyazensis occurs in Bolivia and Brazil (Tropicos, 2014). In Brazil, it is distributed in the Federal District and in the states of Amapà, Amazonas, Espirito Santo, Goiàs, Maranhao, Mato Grosso, Mato Grosso do Sul, Minas Gerais, Parà, Pernambuco, Rondônia, Sergipe and Tocantins, in the Central-West, Northeast, North and Southeast regions and in the Amazon and Cerrado Biomes. It grows in Ciliary Forest or Gallery Forest, Terra Firme Forest, Vàrzea Forest and Semideciduous Dry Forest (Barros *et al.*, 2015). In the Federal District it occurs in mesophytic hillside forest (Mata Seca Decidua) and gallery forest, flowering from July to October, and fruiting from October to December.

Selected material: **APA de Cafuringa,** Fazenda Palestina, X.1992, *Pereira, B.A.S. 2271* (CEN, IBGE). **FERCAL,** Morro da pedreira, 15°30'40"S, 47°57'36"W, XI.2014, *Queiroz, V.V. 15* (UB). **Right bank of the Preto River**, 16°02'S, 47°19'W, X.2002, *Rezende, J.M. 662* (CEN).

Additional material examined: GOIAS. **Luziânia,** Fazenda do Sr. Silas, right bank of the Corumbâ river, VI.2003, *Pereira-Silva, G. 7769* (CEN). **Formosa,** Salto do Itiquira, 15°21'59.8"S, 47°27'13.5"W, XII.2014, *Queiroz, V.V. et al., 18* (UB). **Paraùna,** Ponte de Pedra, IV.1992, *Batista, J.A.N. 300* (CEN).

8. *Lophiaris* Raf., *Fl. Tellur.,* 4, 40 (1836).

Epiphytic **herbs.** Short **rhizome** between the secondary stems. Secondary stems swollen into **pseudobulbs,** very short, subcylindrical, topped by a leaf. **Leaves** conduplicate, bifacial, fleshy, articulated, green, usually spotted with purple. Lateral **inflorescence,** raceme or panicle type, longer than the leaves, multiflora. Resupinate **flowers,** small to large, not very to very showy, with a combination of yellow, white, purple and brown colors, usually spotted. **Lateral sepals** somewhat connate, dorsal sepal and petals free, similar to each other, reticulate venation, papillose on the outer surfaces. **Lip** trilobed, ecalcarate, base of lip inserted perpendicularly to base of column; lateral lobes small in relation to median lobe; disc region with complex callus (1-2 series of teeth, central keel). **Column** winged, wings or stellids flanking the stigma; presence of infra-stigmatic tubule; clinandrium poorly developed, smooth. **Fruit** ellipsoid capsule, not to slightly 3-angled; ripe **fruit** with apex of carpels united.

A genus established by the botanist Constantine Samuel Rafinesque in 1836, but in the *Obra princeps* no reference is made to the etymology, but by deduction possibly from the Greek *lophos* crista and, - *aris* a suffix meaning "provided with", probably an allusion to the series of teeth on the callus that resemble a crest.

The genus has around 36 species distributed from Florida, Mexico, Central America to Argentina and Brazil (WCSP, 2014). Six species occur in Brazil (one endemic), in the Amazon, Cerrado and Atlantic Forest biomes. Two species are represented in the Cerrado biome and of these only *Lophiaris pumila* (Lindl.) Braem occurs in the Federal District (Barros *et al.,* 2015).

Lophiaris species have predominantly yellow or pink flowers. Those with yellow flowers have brownish spots and calluses typical of flowers that produce oil or resin as a reward and are probably visited by bees from the *Centris* and *Paratetrapedia* genera (Silvera, 2002); Parra-Tabla *et al.* (2000) comment that *Lophiaris* flowers are also pollinated by bees from the *Trigona* genera and that some species are pollinated by mistake. Pansarin & Pansarin (2011) report that although the lamellated callus of *Trichocentrum pumilum* Lindl. (= *Lophiaris pumila*) resembles an elaiophore, it does not produce oil as a reward to pollinators. On the other hand, the lateral lobes have elongated trichomes that produce and store lipoid substances, which are responsible for attracting and pollinating two species of bees (*Tetrapedia diversipes* and *Lophopedia nigrispinis*).

8.1. ***Lophiarispumila*** (Lindl.) Braem. Schlechteriana 4: 21 (1993).

Fig. 5. A

Green **pseudobulbs**, 0.2-0.8 x 0.2-0.3 cm, laterally protected by palea sheaths. **Leaves** sessile, obovate to elliptic, bases attenuated, apices acute, light green, 4.4-16.5 cm. Panicle-like **inflorescence**, 60-150 or more flowers, 7.0-23.5 cm long overall. **Flowers** short-pedicellate, pedicels 1.0-2.0 mm. **Sepals** similarly oval-oblong, bases attenuate, margins entire, apices obtuse, membranaceous, burnt-yellow, dorsal sepal 1.0-2.0 x 0.6-1.5 mm; lateral sepals 1\3 adnate, 1.5-2.5 x 0.6-1.5 mm. **Petals** elliptic to subspatulate, bases attenuate, margins entire, apices obtuse, membranaceous, burnt yellow, 1.0-2.1 x 0.8-1.1 mm. **Lip** whitish to burnt yellow, 2.0-4.0 x 0.8-1.1 mm; base slightly cordate; lateral lobes triangular-ovate, burnt yellow, 1.3-1.9 x 1.5-2.1 mm; median lobe obovate, apex entire, obtuse, yellow, 0.5-1.7 x 0.8-1.1 mm; disc region with callus 4-carinate. **Column** 1.0-1.1 x 0.4-0.5 mm, brown, infra-stigmatic tubule 0.5 x 0.4-0.5 mm, wings or stellidia 0.3-0.5 x 0.4-0.6 mm. **Fruit** 8.0 x 5.0-6.0 mm, oblong to ovoid.

Lophiaris pumila occurs in Argentina, Bolivia and Brazil (Tropicos, 2014). In Brazil, it is distributed in the Federal District and the states of Bahia, Espirito Santo, Goiás, Mato Grosso, Mato Grosso do Sul, Minas Gerais, Parâ, Paranâ, Rio de Janeiro, Rio Grande do Sul, Santa Catarina, São Paulo and Sergipe, in the Midwest, Northeast, North, Southeast and South regions, and in the Amazon, Caatinga, Cerrado and Atlantic Rainforest biomes. It grows in Ciliary Forest or Gallery Forest, Terra Firme Forest, Deciduous Dry Forest, Semideciduous Dry Forest, Ombrophilous Forest (= Rain Forest), Mixed Ombrophilous Forest, Restinga (Barros *et al.,* 2015). In the Federal District, it grows in riparian forest, flooded gallery forest, mesophytic deciduous forest and semi-deciduous forest with limestone outcrops. It flowers from November to May and bears fruit from February to May.

Selected material: **APA de Cafuringa,** Fazenda dois irmaos, XII.2002, *Pellizzaro, K.F. 33* (CEN). **FERCAL road,** near the Rio Maranhao bridge, IV.1979, *Bianchetti, L. 414* (HEPH). **Córrego Cabeça de Veado,** 15°52'S, 47°51'O, XII.1993, *Ramos, A.E. 577* (HEPH). **DF-100 direction BSB-Formosa,** Fazenda Manga, 15°57'S, 47°22'O, XI.2002, *Santos, A.A. 1698* (CEN). **FERCAL,** Forest to the right of DF- 205, I.1992, *Batista, J.A.N. 261* (CEN). **Jardim Botànico de Brasilia,** Portion above the leisure area, 15°52'S, 47°51' O, V.1999, *Ramos, A.E. 1301* (HEPH). **Reserva Ecològica do Jardim Botànico de Brasilia,** Córrego Cabeça de Veado gallery forest, 11.2003, *Pellizzaro, K.F. 54* (CEN). **Santuàrio Ecológico do Riacho Fundo,** 15°51'S, 47°57'O, II.1995, *Oliveira, R.S. 64* (UB). **Sobradinho dos Melos,** Fazenda Sr. Benedito de Jesus A. Reis, 15°47'S, 47°43'O, V.2005, *Salles, A.E.H. 3761* (HEPH).

9. *Macroclinium* Barb.Rodr., *Gen. Sp. Orch. Nov.*, 2, 236 (1882).

Epiphytic **herbs.** Short **rhizome** between secondary stems. Secondary stems swollen or not in **pseudobulbs**, if present then reduced, ancipitate, topped by 1-leaf. **Leaves** ensiform, flabelliform, unifacial, fleshy, articulated, green, surface usually reticulate or rough. Sub-umbellate raceme-like **inflorescences** (with congested flowers at the apex), longer than the leaves, multiflora. **Flowers** usually resupinate, small, usually colored white, pink or purple. Free **sepals** and **petals**, similar in size and shape.

Lip entire to slightly trilobed, spatulate to hastate, ecalcarate, base of lip inserted perpendicularly at base of column, base far-unguiculate, terete; disc region with simple, raised callus. **Spine** not winged, slightly reflexed, absence of infrasigmatic tubule, clinandrium poorly developed, smooth. **Fruit** ellipsoid capsule, not to slightly 3-angled; ripe fruit with apex of carpels united.

A genus established by the Brazilian botanist Barbosa Rodrigues, in 1882, in allusion to the large rostellum observed in the type species (Chase, 2009).

The genus has around 40 species with a Neotropical distribution, from Mexico through Central America (except the Antilles) to Peru and Brazil (Chase, 2009). In Brazil there are five species (three endemic) in the Amazon and Cerrado biomes. Only one species occurs in the Cerrado biome, *Macroclinium wullschlaegelianum* (Focke) Dodson, which is present in the Federal District (Barros *et al.,* 2015).

Macroclinium species have delicate flowers with sepals, petals and narrow lips. Despite their small size, the flowers are fragrant and attract relatively large bees (Neubig *et al.*, 2012). *Macroclinium* species do not produce nectar and the pollinators are the same as those observed for *Notylia* species, male Euglossini bees, which collect floral fragrances (Chase, 2009).

9.1. *Macroclinium wullschlaegelianum* (H.Focke) Dodson. Icon. Pl. Trop. 10: t. 939 (1984).

Fig. 2. E-F; **5.** A

Elliptical **pseudobulbs**, greenish-brown, 3.0-7.0 x 1.0-3.0 mm, without sheaths. **Leaves** sessile, elliptic, bases truncated, apices acute, green, 0.7-3.4 x 0.2-0.5 cm. **Inflorescence** 3-10 flowers, 3.0-7.5 cm long overall. **Flowers** pedicellate, pedicels 0.2-1.4 cm. **Sepals** similar, elliptic-linear, bases attenuated, margins entire, apices acuminate, membranaceous, yellowish (cream), dorsal sepal 8.0-11.0 x 1.02.0 mm; lateral sepal free, 9.0-10.0 x 0.1 mm. **Petals** elliptic-linear, bases attenuated, margins entire, apices acuminate, membranaceous, yellowish purple spotted, 7.0-9.0 x 1.0 mm. **Lip** purple, 7.0-9.0 x 2.0-3.0 mm; lateral lobes (auricles) very poorly developed triangular with small callus between them, median region narrowly oblong, 2.0-3.0 mm long, base unguiculate 2.5 x 0.3-0.4 mm, unguicle 1.0 x 0.1-0.2 mm, apical region cordate-triangular, margins slightly erose, 4.0-6.0 x 2.0-3.0 mm. **Column** 9.0-10.0 x 0.2-0.3 mm, brown. **Fruit** 8.0 x 5.0 mm, oblong.

Macroclinium wullschlaegelianum occurs in Colombia, French Guiana, Guyana, Panama, Peru, Suriname, Venezuela and Brazil (Tropicos, 2014). In Brazil, it is distributed in the Federal District and in the states of Amazonas, Goiàs, Maranhao, Parà and Rondônia, in the Central-West, Northeast and North regions, and in the Amazon and Cerrado biomes. It grows in Ciliary Forest or Gallery Forest, Igapó Forest, Terra Firme Forest and Vàrzea Forest (Barros *et al.,* 2015). In the Federal District, it occurs in flooded gallery forest environments and flowers from February to April, with fruiting from April to August.

Selected material: Forest at **DF-100** junction, Ribeirao Jacaré, IV.1992, *Bianchetti, L. 1240* (CEN). **DF-130,** Rio Maranhao, II.1999, *Batista, J.A.N. 878* (CEN). **Fazenda dos Guimaraes,** near CPAC/EMBRAPA, УШ.1991, *Batista, J.A.N. 205* (CEN). **Marginals of the Vicente Pires River,** II.1961, *Heringer, E.P. 7898* (UB).

Additional material examined: GOIAS. **Formosa,** Right bank of the Bezerra River, III.2002, *Pereira-Silva, G. 5973* (CEN). **Luziânia,** Fazenda Suindara do Alagado, XI.2002, *Pereira-Silva, G. 6947* (CEN).

10. *Notylia* Lindl., *Bot. Reg.,* 11, sub. T. 930 (1825).

Epiphytic **herbs.** Short **rhizome** between secondary stems. Secondary stems swollen into **pseudobulbs**, spherical to narrowly elliptic, topped by 1-2 leaves. **Leaves** bifacial, conduplicate, leathery, articulated, green. **Inflorescence** lateral, raceme-like, pendulous, longer than the leaves, multiflora. Small, cream, white, yellow or green resupinate **flowers,** sometimes with yellow markings on the petals. **Sepals and petals** similar, dorsal sepal free, lateral sepals partially fused.

Lip entire, hastened, echolocate, base of lip inserted perpendicularly at base of column, base unguiculate, terete; disc region with simple callus, usually elongated, carinate. **Spine** not winged, slightly reflexed; absence of infrasigmatic tubule; clinandrium poorly developed, smooth. **Fruit** ellipsoid to suborbicular capsule, not to slightly 3-angled; mature fruit with apex of carpels united.

A genus established by the English botanist John Lindley, in reference to the unique recurved apical portion of the spine (Chase, 2009).

The genus has around 60 species, distributed from Mexico to Central America, Panama, Colombia, Venezuela, Bolivia, Paraguay, Peru and Brazil (Chase, 2009). In Brazil there are 25 species (15 endemic), in the Amazon, Cerrado and Atlantic Forest biomes. There are 10 species in the Cerrado biome and two in the Federal District, *Notylia hemitricha* Barb.Rodr. and *Notylia lyrata* S.Moore (Barros *et al.*, 2015).

Unlike the species of the closest genus, *Macroclinium, Notylia* species have pendulous racemes with flatter flowers. Pollination is also carried out by male Euglossini bees that collect floral fragrances (Chase, 2009; Neubig *et al.*, 2012).

10.1. *Notylia lyrata* S.Moore. Trans. Linn. Soc. London, Bot. 4: 477 (1895).

Fig. 2. G; **5.** B.

Green **pseudobulbs**, 0.9-3.0 x 0.2-0.6 cm, laterally protected by persistent, articulated leaf sheaths, 3.0-8.1 cm, topped by 1 leaf. **Leaves** sessile, elliptic to narrowly obovate, bases attenuated, apices obtuse, 4.4-12.0 x 0.8-2.6 cm. **Inflorescence** 10-90 or more flowers, 5.0-20.0 cm long overall. **Flowers** pedicellate, pedicels 3.0-5.0 mm long. **Sepals** similar, narrowly elliptic to linear, bases attenuate, margins entire, apices acute, membranaceous, yellow, dorsal sepal 4.0-5.0 x 0.6-1.5 mm, lateral sepals 2/3 adnate, 3.0-6.5 x 0.3-1.0 mm. **Petals** elliptic to linear, bases attenuated, margins entire, apices acute, membranaceous, yellow-white with two orange basal spots, 3.5-6.0 x 0.7-1.1 mm. **Lip** yellowish-white, 3.0-5.0 x 1.4-2.5 mm; base unguiculate, apex acute to obtuse, yellowish, 0.3-0.8 x 0.2-0.5 mm, mid-apical lobe oval-lanceolate, whitish, 2.5-4.0 x 1.5-2.5 mm; disc region with fleshy callus, elongated, carinate. **Column** 1.5-2.5 x 0.3-1.0 mm, brown. **Fruit** 8.0-9.0 x 4.0-5.0 mm, oblong.

Notylia lyrata occurs in Paraguay and Brazil (WCSP, 2014). In Brazil, it is distributed in the Federal District and in the states of Amazonas, Bahia, Cearà, Espirito Santo, Goiàs, Maranhao, Mato Grosso, Minas Gerais, Parà, Paraiba, Paranà, Pernambuco, Rio de Janeiro, Rio Grande do Sul, Santa Catarina, Sao Paulo and Tocantins, mainly in the Amazon, Cerrado and Atlantic Rainforest biomes. It grows in Ciliary Forest or Gallery Forest, Terra Firme Forest, Dry Semideciduous Forest, Ombrophilous Forest (= Rain Forest) and Restinga (Barros *et al.*, 2015). In the Federal District, it grows in riparian forest and gallery forest, flowering from June to September and fruiting from October to November.

Selected material: **BR-020,** Fazenda Rio Preto, XI.1982, *Ramos, A.E. 166* (CEN). **FERCAL road,** near the Maranhao River, VIII.1979, *Bianchetti, L. 432* (HEPH). **Fazenda Agua Limpa,** UnB, 15°50'S, 48°00'O, VI.1988, *CRGF Group, s.n.* (CEN15122). **IBGE,** Córrego Roncador, 15°57'S, 47°52'O, VIII.1989, *Azevedo, M.L.M. 297* (IBGE). **Nùcleo Rural Jardim II,** Right bank of the Rio Preto, 16°00'S, 47°22'O, X.2002, *Rezende, J.M. 630* (CEN). **Parque Nacional de Brasilia,** Mata córrego do acampamento, 15°45'39"S, 47°58'40"W, VI.2014, *Queiroz, V.V. & Lima, J.H. 13* (UB). **Santuàrio Ecològico do Riacho Fundo,** 15°51' S, 47°57'O, X.1994, *Oliveira, R.S. 8* (UB).

Notylia hemitricha Barb.Rodr. is cited for the Federal District by the List of Species of the Flora of Brazil (Barros *et al.,* 2015). The testimonial materials cited by this list [Ghillany, A., s.n., <u>HB</u> 68909; Klein, R.M., Bresolin, A., 8585, <u>FLOR</u>, (SC); Fraga, C.N., 127, <u>MBML</u>, (ES)], are not from the Federal District. As no material witnessing the supposed occurrence has been located, we believe this to be a misidentification when we consider the morphological similarity between *N. hemitricha* and *N. lyrata* (which does occur).

11. *Plectrophora* H. Focke, *Tijdschr. Nat. Wetensch.,* 1, 212 (1848).

Epiphytic **herbs.** Short **rhizome** between secondary stems. Secondary stems swollen or not in **pseudobulbs,** if present then reduced, ancipitate, topped by 1-leaf. **Leaves** ensiform, flabelliform, unifacial, fleshy, smooth-surfaced, articulated, green. Lateral, raceme-like **inflorescence,** smaller than the leaves, usually uniflorous. **Flowers** resupinate, not very showy, greenish, white or yellow, sometimes with a reddish-orange spot on the lip. Dorsal **sepals** free, the lateral ones fused at the base involving a nectariferous spur. **Petals** free. **Lip** entire to slightly bilobed, margins curved over the column, calcarate, base of the lip inserted perpendicularly at the base of the column, base prolonged forming a spur protected by the lateral sepals; disc region flared or with a pair of ridges at the base. **Column** not winged, weakly arched; absence of infrasigmatic tubule; clinandrium poorly developed, smooth. **Fruit** ellipsoid capsule, 3-winged to strongly 3-angled; mature fruit with apex of carpels united.

A genus established by the Dutch botanist Hendrik Charles Focke in 1848, in reference to the presence of a spur in the flowers (Chase, 2009).

The genus has around nine species distributed in Costa Rica, Panama, Venezuela, Colombia, Ecuador, Peru Bolivia and Brazil (Chase, 2009). In Brazil there are five species (three endemic) in the Amazon and Cerrado biomes. In the Cerrado biome there are three species and in the Federal District there is one species, *Plectrophora edwallii* Cogn. (Barros *et al.,* 2015).

There is no information on pollination mechanisms or pollinators for *Plectrophora* species. However, *Plectrophora* species are small plants with large flowers and have a funnel-shaped lip and a spur formed by the lateral sepals, but without nectar-producing glands (Neubig *et al.*, 2012). Chase (2009), citing Dodson (2003), considers the relative similarity of *Plectrophora* flowers to *Trichocentrum s.s.* flowers and believes that these two genera have the same pollination syndrome, i.e. he intuits that *Plectrophora* may be pollinated by male Euglossini bees that collect fragrances like *Trichocentrum*. According to Neubig *et al.* (2012), the flowers are probably pollinated by mistake, by insects that forage for nectar and have long tongues.

11.1. *Plectrophora edwallii* Cogn. Fl. Bras. 3(6): 580 (1906).

Fig. 2. C-D; **5.** B.

Pseudobulbs broadly elliptic to suborbicular, extremely ancipitate, yellowish green, 9.0 x 7.0 mm, protected by articulated, persistent sheaths, 4.9-8.0 cm. **Leaves** elliptic to narrowly oval, base truncated, apex acute, green, 6.2-12.4 x 0.6-1.0 cm. **Inflorescence** short, up to 3 flowers opening in succession, 4.9 cm long overall; **flowers** pedicellate, pedicels 1.1-1.9 cm. **Sepals** similar, oval, bases attenuated, margins entire, apices acute, membranaceous, pale yellow, dorsal sepal 10.0 x 3.0 mm; lateral sepals 8.0-10.0 x 2.0 mm. **Petals** elliptic, bases attenuated, margins entire, apices acute, membranaceous, yellow, 11.0 x 5.0 mm. **Lip** obovate, margin erose, apex obtuse, yellow with darker streaks along the veins, 13.0 x 8.0 cm; spur 9.0 x 1.0-2.0 mm. **Column** 3.0 x 1.0 mm. **Fruit** 2.0 x 1.5 cm, ellipsoid.

Plectrophora edwallii occurs in Bolivia, Peru and Brazil (Tropicos, 2014). In Brazil, it is distributed in the Federal District and the states of Goiàs, Mato Grosso, Minas Gerais and Parà, in the Central West, North and Southeast regions, and in the Amazon and Cerrado biomes. It grows in Ciliary Forest or Gallery Forest, Terra Firme Forest, Semideciduous Dry Forest (Barros *et al.*, 2015). In the Federal District, it grows in Ciliary Forest on limestone outcrops, Mesophytic Semideciduous Forest (Mata Seca Semidecidua), and flowers in April and August, and bears fruit from April onwards.

Specimen: **FERCAL,** Entrance to Pedra Encantada cave, IV.2006, *Rezende, J.M. 1071* (CEN). Margens do **Rio Preto,** Divisa do DF com MG, VIII.1963, *Heringer, E.P. 9186* (UB).

Additional material examined: GOIAS. **Formosa,** Salto do Itiquira, 15°22'02.6"S, 47°27'10.8"W, XII.2014, *Queiroz, V.V. et al., 22* (UB, CEN). **Niquelândia,** Areas on the final stretch of the Tocantinzinho river, XI.1996, *Walter, B.M.T. 3615* (CEN).

12. *Rodriguezia* Ruiz & Pav., *Fl. Peruv. et Chil. Prodr.,* 115, t. 25 (1794).

Epiphytic **herbs.** Short to extremely long **rhizome** between the secondary stems. Secondary stems swollen into **pseudobulbs**, elongated to orbicular, topped by 14 leaves. **Leaves are** bifacial, conduplicate, leathery, articulated, green. Lateral **inflorescence,** usually raceme-like, usually larger than the leaves, pauci to multiflora. Showy, yellow, pink, carmine **flowers**, often with red and yellow markings on the sepals and lip. Dorsal sepal free; base of lateral sepals fused around nectariferous mentum. **Petals** free. **Lip** entire, oblanceolate, base of lip inserted perpendicularly to the base of the column, base prolonged forming a small mentum or not, apex dilated, usually emarginate; disc region with one or more carinae or ridges. **Column** with a pair of developed arms at the apex and a pair of smaller arms flanking the stigma, slightly projected base secreting nectar inside the mentum, absence of infrasigmatic tubule, clinandrium poorly developed, smooth. **Fruit** elliptic to ovoid capsule, not to slightly 3-angled; mature fruit with apex of carpels united.

A genus established by the Spanish botanists Hipólito Ruiz López and José Antonio Pavón Jiménez in 1794 in homage to Antonio Manuel Rodriguez de Vera (1780-1846), a Spanish botanist and pharmacist at the Spanish court (Chase, 2009).

The genus has around 48 species distributed from Mexico, Central America, Colombia, Venezuela, Argentina, Bolivia, Peru and Brazil (Chase, 2009). In Brazil there are around 23 species (17 endemic) in the Amazon, Cerrado and Atlantic Forest biomes. Four species occur in the Cerrado biome and *Rodriguezia decora* (Lem.) Rchb.f. occurs in the Federal District.

Rodriguezia species have showy, relatively large and brightly colored flowers, unlike the other "twig" species, which remain on the treetops in the outer layer and receive high levels of light and radiation. The lip is often relatively large and flat and the lateral sepals are fused to form an elongated nectary spur or just a small mentum. In some species, the two projections at the base of the column secrete nectar into the spur. Neubig *et al.* (2012), based on the work of Chase (2009), point out that the Brazilian species *Rodriguezia decora* does not fit the characteristics shared by most species of the genus, such as a long rhizome between the pseudobulbs and the absence of an elongated spur. According to Dodson (1965), for the species of *Rodriguezia* that produce nectar, except for *R. decora,* hummingbirds, butterflies and bees that forage in search of nectar are recorded as pollinators.

12.1. *Rodriguezia decora* **Rchb. f. Bot. Zeitung (Berlin) 10: 771 (1852).**

Fig. 3. A-C; **5.** C

Long **rhizome** between the pseudobulbs, 3.0-19.5 cm. **Pseudobulbs** elliptic to obovate, greenish-brown, 1.0-2.5 x 0.4-1.7 cm, laterally protected by articulated, persistent leaf sheaths, 4.2-12.7 cm, topped by 1 leaf. **Leaves** sessile, narrowly ovate to elliptic, base attenuated, apex acute, rounded to mucronate, leathery, 2.9-13.0 x 0.7-2.3 cm. **Inflorescence** 5-15 or more flowers, 12.0-29.0 cm long overall. **Flowers** pedicellate, pedicels 0.8-1.1 cm long. **Sepals** similar, elliptic, bases attenuated, margins entire, apices acute, membranaceous, white or slightly pinkish with a wine tinge, dorsal sepal 0.8-1.7 x 0.3-0.5 cm, lateral sepals 2/3 adnate, 1.3-1.9 x 0.1-0.3 cm, covering a small mentum at the base of the lip. **Petals** oval, bases attenuated, margins entire, apices obtuse, membranaceous, wine-white with a slight pinkish tinge in the center, 1.3-1.8 x 0.3-0.8 cm. **Lip** base attenuated, unguiculate, presence of small mentum, median part spotted with burgundy, apical part dilated, rounded-reniform, apex emarginate, white without spots, 1.6-3.0 x 1.0-1.9 cm; disc region 2-lamellate, serrate. **Column** 4.0-7.0 x 1.0 mm, dark brown, 4 apical arms, ligulate, posterior arms smaller, 1.0-2.0 mm long, anterior arms larger, 3.0-4.0 mm long. **Fruit** not seen.

Rodriguezia decora occurs in Paraguay, Argentina and Brazil (WCSP, 2014). In Brazil, it is distributed in the Federal District and in the states of Paranâ, Rio Grande do Sul, Santa Catarina and São Paulo, in the Central West, Southeast and South regions, and in the Cerrado and Atlantic Forest biomes. It grows in Riparian Forest or Gallery Forest, Deciduous Dry Forest, Semideciduous Dry Forest, Ombrophilous Forest (= Rain Forest) (Barros *et al.,* 2015). In the Federal District, it grows in Flooded Gallery Forest environments and flowers from December to March.

Selected material: **Colégio Agricola,** XII.2003, *Salles, A.E.H. et al., 2899* (HEPH). **Fazenda Sucupira,** Mata do Riacho Fundo, III.2000, *Batista, J.A.N. 1052* (CEN). Forest **between Guarà and the Industry Sector,** II.1981, *Chagas, F. 369* (IBGE). **Brasilia National Park,** Córrego do Acampamento, III.1983, *Ramos, A.E. 234* (CEN, HEPH). **Zoobotanical Park,** III.1961, *Heringer, E.P. 8063* (UB). **Guarà Ecological Reserve,** II.2001, *Miranda, Z.J.G. 71* (CEN). **Reserva Ecològica do IBGE,** Córrego Taquara, 15°56'S, 47°55'O, III.1994, *Walter, B.M.T. 2089* (CEN).

Rodriguezia brachystachys Rchb.f. & Warm is also cited for the Federal District (Barros *et al.,* 2015), but no *voucher* was found for the region and for this reason it was removed from the update.

Rodriguezia decora var. *lactea* L.C.Menezes, is also cited for the Federal District (Menezes 1995). Examination of the type material (Brasilia, Mata Alagada do Guarà. No date. *Menezes, L.C. 36,* UB) showed that there were no flowers or inflorescences. According to information taken from the Lista das Espécies da Flora do Brasil (2015*) Rodriguezia decora* var. *lactea* L.C.Menezes is a synonym of *Rodriguezia decora* Rchb. f.

Just to illustrate the infraspecific category, Barros & Batista (2004) consider that this category would be better treated as a form and not as a variety (*Rodriguezia decora* f. *lactea* (L.C.Menezes) F.Barros & J.A.N.Bat.).

13. *Sanderella* Kuntze, *Rev. Gen. Pl.,* 2, 649 (1891).

Epiphytic **herbs.** Short **rhizome** between secondary stems. Secondary stems swollen into

pseudobulbs, elliptic, ancipitate, topped by 1 leaf. **Leaves** conduplicate, bifacial, leathery, articulated, green or purplish. Lateral, raceme-like **inflorescence,** smaller or the same length as the leaves, multiflora. Small, inconspicuous, greenish-white **flowers** with reddish-brown spots. Dorsal sepal free; lateral sepals mostly fused. **Petals** free. **Lip** entire, oblong to pandurate, ecalcarate, base of lip inserted perpendicularly to base of column, apex not emarginate; disc region with raised callus. **Column** winged, small projections flanking the stigma, apex with discrete arms projecting forward, absence of infrasigmatic tubule, clinandrium poorly developed, smooth. **Fruit** ellipsoid capsule, not to slightly 3-angled; mature fruit with apex of carpels united.

A genus established by the German botanist Carl Ernst Otto Kuntze in 1891 in homage to the British orchid specialist Henry F. C. Sander (1847-1920), founder of F. Sander & Company and author of the first list of orchid hybrids (Chase, 2009).

The genus has two species distributed in Bolivia, Argentina and Brazil. In Brazil, the species occurs in the Cerrado and Atlantic Forest biomes (Chase, 2009). Only *Sanderella discolor (*Barb.Rodr.) Cogn. occurs in the Cerrado biome and the Federal District (Barros *et al.,* 2015).

Singer and Cocucci (1999) reported visits from bees and wasps in *Capanemia* and commented that Càssio van den Berg (in a personal communication) made reference to the morphological similarity between *Sanderella* and *Capanemia* and the possibility that the two species have the same pollinators.

13.1. *Sanderella discolor* (Barb.Rodr.) Cogn. Fl. Bras. 3(6): 239 (1905).

Fig. 5-C.

Green **pseudobulbs**, 5.0-12.0 x 1.0-5.0 mm, laterally protected by palea sheaths, persistent, 0.6-2.7 cm. **Leaves** sessile, elliptic to lanceolate, bases attenuated, apices acute, leathery, dark green, 2.4-7.7 x 0.3-1.0 cm. **Inflorescence** 5-19 flowers, 1.5-8 cm long overall. **Flowers** pedicellate, pedicels 1.0-4.0 mm. **Sepals** similar, slightly obovate, narrowed bases, entire margins, acute apices, membranaceous, whitish with vinaceous bases, dorsal sepal 2.0-3.0 x 1.0-2.0 mm, lateral sepal 2\3 adnate, 2.0-3.0 x 1.0-2.0 mm. **Petals** obovate, bases truncate, margins entire, apices acute, membranaceous, whitish, 2.0-3.0 x 1.0 mm. **Lip** pinnate, apex obtuse, reflexed, slightly acuminate, 2.0-3.0 x 0.6-1.0 mm; disc region with bilamellar callus. **Column** 1.0-1.5 x 0.3-0.7 mm, brown. **Fruit** not seen.

Sanderella discolor occurs in Argentina, Bolivia and Brazil (Tropicos, 2014). In Brazil, it is distributed in the Federal District and in the states of Goiàs, Minas Gerais, Paranà, Rio Grande do Sul and Sao Paulo, in the Midwest, Southeast and South regions, in the Cerrado and Atlantic Forest biomes. It grows in phytophysiognomies that generally have a high level of humidity, such as

Riparian Forest or Gallery Forest, Dry Semideciduous Forest, Ombrophilous Forest (= Rain Forest) (Barros *et al.,* 2015). In the Federal District, it occurs in riparian forest, flooded gallery forest, and blooms in May and June.

Selected material: **Córrego Acampamento,** V.1991, *Batista, J.A.N. 199* (CEN). **Fazenda Sucupira,** VI.1995, *Assis, M.C. 230* (CEN). **Margem do Riacho Fundo,** Fundaçao Zoobotânica, n.d., *Heringer, E.P. s.n.* (UB8333). **Guarà Ecological Reserve,** XI.2003, *Batista, J.A.N. 1454* (CEN). **Sobradinho dos Melos,** VI.2005, *Salles, A.H. 3787* (HEPH).

14. *Trichocentrum* Poepp. & Endl., *Nov. Gen. Spec. Pl.,* 2, 11, t. 115 (1836). *"Strictu sensu"*

Epiphytic **herbs.** Short **rhizome** between secondary stems. Secondary stems swollen into small (reduced) **pseudobulbs,** ovoid to cylindrical, topped by 1 leaf. **Leaves** conduplicate, bifacial, fleshy, articulated, green, usually spotted with purple. Lateral **inflorescence,** raceme or panicle type, shorter than the leaves, pauciflorous. Showy, resupinated **flowers in a** combination of brown, pink, purple, yellow and white. **Sepals** and **petals** free and similar to each other. **Lip** entire to three-lobed, calcarate, base of lip inserted perpendicular to base of column, apical lobe bilobed, margins entire; disc region with tuberculous callus or with longitudinal carinae or blades. **Column** not winged, auricles or apical arms developed, petaloid, absence of infrasigmatic tubule, clinandrium poorly developed, smooth. **Fruit** ellipsoid capsule, usually not 3-angled, occasionally winged; mature fruit with apex of carpels united.

A genus established by the German botanist Eduard Friedrich Poeppig and the Hungarian botanist Stephan Friedrich Ladislaus Endlicher in 1836, in reference to the presence of trichomes or papillae on the inner walls of the spurge (Pupulin, 1995; Chase, 2009).

The genus has around 30 species distributed throughout the Neotropics, from southern Florida, Mexico, Central America, the Caribbean, Argentina, Uruguay, Paraguay, Peru and Brazil (Pupulin, 1995; Chase, 2009). In Brazil there are five species (two endemic) in the Amazon, Cerrado and Atlantic Forest biomes. Two species occur in the Cerrado and only *Trichocentrum albococcineum* Linden occurs in the Federal District (Barros *et al.,* 2015).

Chase (2009) comments on the reports by van der Pijl & Dodson (1966) and Roubik & Ackerman (1987) about *Trichocentrum s.s.* species having the Euglossini bee pollination syndrome. In many cases, *Trichocentrum* species have a non-functional nectary, i.e. a spur without nectar, and this may be associated with two pollination strategies: attracting Euglossini males that collect fragrance, and females that forage in search of nectar and are deceived by the presence of the spur.

14.1. *Trichocentrum albococcineum* Linden. Ann. Hort. Belge Étrangère 15: 103 (1865).

Fig. 3. G-H; **5.** D

Green **pseudobulbs**, 0.5-0.7 x 0.2-0.4 cm. **Leaves** sessile, elliptic to ovate, base attenuated, apex acute, 6.4-13.6 x 1.2-1.8 cm. **Inflorescence** raceme-like, 1-2 flowers, 4.0-4.5 cm long overall. **Flowers** pedicellate, pedicels 3.0-4.0 cm. **Sepals** obovate, base slightly narrowed, apex acute-acuminate, membranous, brownish-green, dorsal sepal 1.9 x 0.9 cm, lateral sepal 1.9-2.2 x 0.5-0.7 cm; **petals** obtuse apex, 1.8-1.9 x 0.7-0.9 cm. **Lip** pinnate, white with two coccine lateral bands, 2.8-3.4 x 2.0-2.2 cm; base narrowed, 9.0 mm long, apex emarginate, spur 6.0-8.0 mm long; disc region with lamellate callus. **Column** 0.7 x 0.4 cm, dark brown. **Fruit** not seen.

Trichocentrum albococcineum occurs in Bolivia, Peru and Brazil (Tropicos, 2014). In Brazil, it is distributed in the Federal District and in the states of Amazonas, Goiàs, Mato Grosso, Mato Grosso do Sul, Parà and Rondônia, in the Central West and North regions, and in the Amazon and Cerrado biomes. It grows in Ciliary Forest or Gallery Forest, Igapó Forest, Terra Firme Forest, Semideciduous Dry Forest (Barros *et al.,* 2015). In the Federal District, it grows in gallery forest and flowers in February.

Examined material: **Aguas Emendadas Ecological Station,** II.2014, *Lima, J.H. 55* (UB).

Additional material examined: GOIAS. **Catalao,** Fazenda Barra, 3 km from Davinopolis, n.d. *Salles, A.H. et al., 2658* (HEPH).

15. *Trichopilia* Lindl. *Bot. Reg.,* 22, t. 1863 (1836).

Epiphytic **herbs.** Short **rhizome** between secondary stems. Secondary stems swollen into **pseudobulbs,** terete to laterally ancipitate, topped by 1 leaf. **Leaves** conduplicate, bifacial, fleshy to leathery, articulated, green. Lateral, raceme-like **inflorescence,** usually shorter than the leaves, pauciflora. Small, resupinate **flowers,** usually white with pink spots. **Sepals** and **petals** free and similar to each other. **Lip** entire, ecalcarate, base of lip inserted perpendicularly to base of column, basal margins curved over column; disc region with carinae near base, sometimes papillose. **Column** half the length of the lip, straight; infra-stigmatic tubule absent; clinandrium very developed (length above the anther), lacerated. **Fruit** elliptic to ovoid capsule, not to slightly 3-angled; mature fruit with apex of carpels united.

A genus established by the English botanist John Lindley in 1836, in reference to the clinandrum, which has a lacerated, felt-like texture, observed in the type species (Chase, 2009).

The genus has around 26 species distributed throughout Mexico, Central America and Venezuela, Colombia, Ecuador, Bolivia, Peru and Brazil (Chase, 2009). In Brazil there are five species (two endemic) in the Amazon, Cerrado and Atlantic Forest biomes. In the Cerrado biome (Barros *et al.,* 2015), and in the Federal District, there is only one species, *Trichopilia brasiliensis* Cogn., which until now had not been cited for the Federal District and was recorded for the first time in this study.

Trichopilia species have the lip forming a tubular structure, with the basal part surrounding the column, suggesting a possible nectar deposit (Neubig *et al.,* 2012). Dodson (1962) reported the pollination of male Euglossini fragrance-collecting bees. Some species of *Cattleya* and *Sobralia* have similar flowers and are visited by male Euglossini bees.

15.1. ***Trichopilia brasiliensis*** Cogn. Fl. Bras. 3(6): 580 (1906).

Fig. 3. D; **5.** D

Pseudobulbs terete to subcylindrical, green, 1.8-2.4 x 0.2-0.3 cm, protected by persistent palea sheaths. **Leaves** sessile, narrowly elliptic to linear, truncate base, acute apex, leathery, 16.0-18.0 x 0.7-0.8 cm. **Inflorescence** 1-5 flowers, 3.06.0 cm long overall. **Flowers** pedicellate, pedicel 0.9-1.1 cm. **Sepals and petals** elliptic, narrowed base, acuminate apex, membranaceous, cream-green with small purplish spots, dorsal sepal 1.5-1.6 x 0.3-0.4 cm, lateral sepal 1.6-1.8 x 0.2-0.3 cm, petals 1.4-1.5 x 0.4 cm. **Lip** broadly elliptic, concave, base with two protuberances flanking the column, apex acute, cream-colored with more purple spots than the sepals and petals and sometimes with dark longitudinal carinae, 1.2-1.3 x 1.0-1.1 cm. **Column** 4.0-5.0 mm, brown. **Fruit** not seen.

It occurs exclusively in Brazil (WCSP, 2014). It is distributed in the Federal District and the states of Espírito Santo, Goiás, Maranhao, Mato Grosso, Pará and Rondônia, in the Central West, Northeast, North, Southeast regions, and in the Amazon, Cerrado and Atlantic Forest biomes. It grows in Riparian Forest, Gallery Forest, Terra Firme Forest, Ombrophilous Forest (= Rain Forest) (Barros *et al.,* 2015). In the Federal District, it grows in riparian forest and flowers in January.

Material examined: **DF-100,** Sentido Formosa, Fazenda Manga, 15°57'S, 47°22'O, I.2007, *Santos, A.A. 1769* (CEN).

Additional material examined: GOIAS. **Pirenópolis,** Santuàrio Ecològico do Vaga Fogo, II.1992, *Bianchetti, L. 1238* (CEN).

5. CONCLUSION

Subtribe Oncidiinae Benth. is well represented in the Federal District, occurring in several protected environments, such as reserves and conservation units, as well as environments not protected by environmental legislation. Based on the results generated by this study, the List of Species of the Flora of Brazil - Federal District (Barros *et al.*, 2015) should be updated as follows: the addition of one more genus and two species, *Trichopilia brasiliensis* Cogn. and *Cohniella jonesiana* (Rchb.f.) Christenson. The inclusion of *Cohniella jonesiana*, even though it does not have a *voucher* for the Federal District, is justified by the fact that it has been cited in previous works (Pellizzaro *et al.*, 2004; Bianchetti *et al.*, 2005) and because it has records in localities close to the Federal District, such as Unai - Minas Gerais, for example. The suppression of *Coppensia bifolia* (Sims) Dumort and *Notylia hemitricha* Barb. Rodr as a result of misidentified samples. Suppression of the occurrence of *Rodriguezia brachystachys* Rchb.f. & Warm and *Gomesafoliosa* (Hook.) Klotzsch & Rchb.f. due to the fact that they do not have *vouchers* or records for the vicinity of the Federal District. Therefore, the representation of the subtribe Oncidiinae for the Flora of the Federal District should be updated to 15 genera and 18 species.

Today, there are two different taxonomic approaches to the subtribe Oncidiinae: a) an approach that prefers the recognition of smaller genera (*Alatiglossum, Cohniella, Coppensia, Gomesa (n.s.), Lophiaris, Trichocentrum* (n.s.)), with a narrower delimitation, with *monophyletic support and easy morphological recognition.*), with a narrower delimitation, monophyletic support and easy morphological recognition which, for the species occurring in the Federal District, form a group of 15 genera and 18 species and, b) an approach which concerns the preference for recognizing larger genera (*Gomesa* (s.l.) and *Trichocentrum (s.l.)*, with a broader or wider delimitation, with monophyletic support, but difficult to recognize morphologically, given the great morphological circumscription which, for the species occurring in the Federal District, form a set of 12 genera and 18 species.

Although both approaches are important and deal with knowledge of the evolution of genera and their circumscription, it is recommended that, in studies dealing with diversity or richness for the Federal District, it is made explicit which approach was used, as the results will be strongly influenced due to the great variation in the number of genera.

6. LIST OF EXSICCATES

Assis, M.C.: 230 (13.1); Augusto, M.M.: 1 (2.1); Azevedo, M.L.M.: 297 (10.1), 506 (5.2); Batista, J.A.N.: 54 (5.1), 118 (5.1), 199 (13.1), 205 (9.1), 261 (8.1), 300 (7.1), 306 (3.1), 337 (5.1), 800 (5.1), 878 (9.1), 935 (6.1), 1052 (12.1), 1101 (3.1), 1454 (13.1), s.n. CEN 60698 (8.1); Belém, R.P.: 1896 (5.1); Bianchetti, L.: 414 (8.1), 424 (13.1), 432 (10.1), 869 (1.2), 869-A (1.2), 869-B (1.1), 1161 (3.1), 1238 (15.1), 1239 (5.2), 1240 (9.1), 1460 (4.1), n.r. CEN 15096 (6.1); Chagas, F.: 369 (12.1); Faria, J.G.: 254 (5.2); Figueiredo, S.: 178-A,B (5.2), 195-A,B (5.2); Fonseca, M.L.: 1031 (1.2); Gomes, S.M.: 86 (5.2); Grupo da CRGF: n.r. CEN 15122 (10.1); Heringer, E.P.: 1831 (5.2),

3744 (5.1), 7380 (12.1), 7898 (9.1), 8063 (12.1), 8065 (5.2), 8447 (5.1), 9186 (11.1), 16751 (4.1), 18358 (11.1), 18462 (11.1), n.r. UB8333 (13.1); Ianhez, M.: 53 (6.1), 56 (6.1); Irwin, H.S.: 10607 (5.1); Lima, J.H.: 16 (4.1), 55 (14.1); Mendes, R.A.: 339 (3.1); Menezes, L.C.: 8 (5.2), 36 (12.1); Miranda, Z.J.G.: 71 (12.1), 82 (1.1); Nobrega, M.G.G.: 432-A,B (5.2), 1294 (6.1); Oliveira, R.S.: 8 (10.1), 16 (4.1), 28 (5.1), 64 (8.1); Pellizzaro, K.F.: 27 (1.2), 28 (6.1), 30 (3.1), 33 (8.1), 54 (8.1); Pereira Neto, M.: 515 (5.1); Pereira, B.A.S.: 2112 (7.1), 2271 (7.1); Pereira-Silva, G.: 5973 (9.1), 6947 (9.1), 7769 (7.1); Pinagé, E.R.: s.n. UB 14529 (5.1); Queiroz, V.V.: 13 (10.1), 15 (7.1), 18 (7.1), 22 (11.1); Ramos, A.E.: 166 (10.1), 234 (12.1), 577 (8.1), 1301 (8.1); Ratter, J.: 4293 (5.1); Rezende, J.M.: 255 (5.1), 630 (10.1), 662 (7.1), 1071 (11.1); Salles, A.E.H.: 3 (5.2), 1730 (5.1), 2658 (14.1), 2899 (12.1), 3761 (8.1), 3787 (13.1), s.n. CEN 026596 (5.1); Santos, A.A.: 398 (4.1), 1618 (2.1), 1698 (8.1), 1769 (15.1); Silva, J.S.: 318 (5.2); Soares, A.C.A.: n.r. HEPH 29268 (5.2); Sucre, D.: 607 (3.1), 647 (1.2); Valente, I.: 19 (4.1); Viana, P.L.: 2826 (3.2); Vieira, R.: 361 (6.1); Walter, B.M.T.: 2089 (12.1), 2091 (5.2), 3615 (11.1), 3774 (1.1).

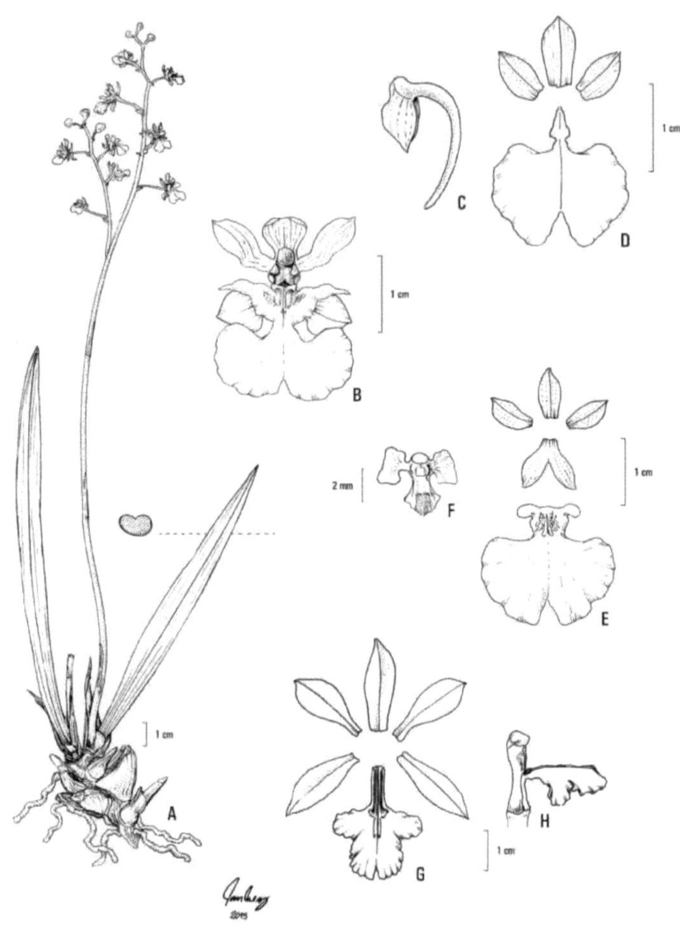

Figura 1. A-B. **Cohniella cepula:** A. Habit; B. Flower; C-D. **Comparettia coccinea:** C. Fused lateral sepals (nectariferous spur); D. Open flower; E-F. **Coppensia hydrophila:** E. Flower open; F. Column, wings or stellidia and infra-stigmatic tubule; G-H. **Aspasia variegata: G.** Flower open; H. Lip inserted in the median part of the column. A-B. (*Pellizzaro 30*; C-D. *Bianchetti 1460;* E-F. *Batista 337*; G-H. *Augusto 1.*)

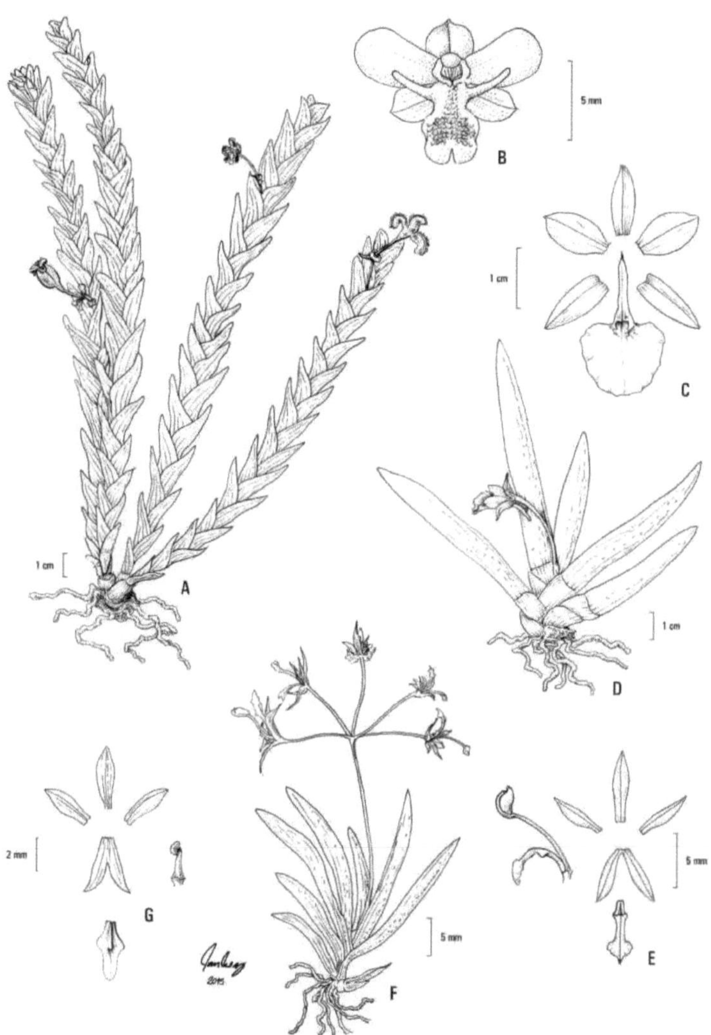

Figura 2. A-B. **Lockhartia goyazensis:** A. Habit; B. Flower; C-D. **Plectrophora edwallii:** C. Open flower; D. Habit; E-F. **Macroclinium Wullschlaegelianum:** E. Open flower and column; F. Habit; G. **Notylia lyrata:** Open flower and column. A-B. (*Pereira 2271\ C-D. Batista 95*; E-F. *Batista 878*; *G. Pereira-Silva 7575.*)

47

Figura 3. A-C. **Rodriguezia decora:** A. Habit; B. Flower, front; C. Flower, lateral, sepals forming a nodule; D. **Trichopilia brasiliensis:** Open flower and column; E-F. **Ionopsis utricularioides:** E. Flower open; F. Lateral sepals, forming small mentum (lateral view); G-H. **Trichocentrum albococcineum: G.** Flower in lateral view, presence of spur; H. Flower. A-C. (*Bianchetti 853*; *Santos 1769*; E-F. Walter *4235*; G-H. *Lima 55.*)

DISTRIBUTION MAP IN THE FEDERAL DISTRICT

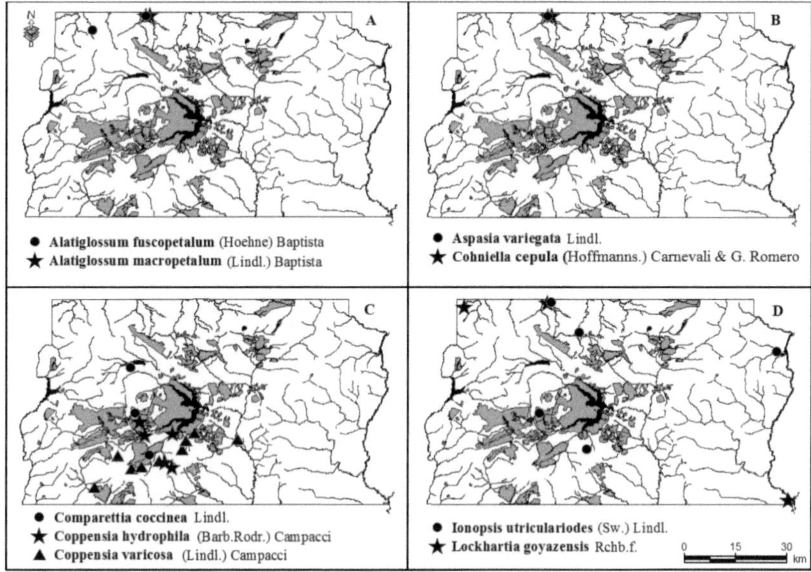

Figura 4.   Distribution of Oncidiinae species in the Federal District, Brazil.

DISTRIBUTION MAP IN THE FEDERAL DISTRICT

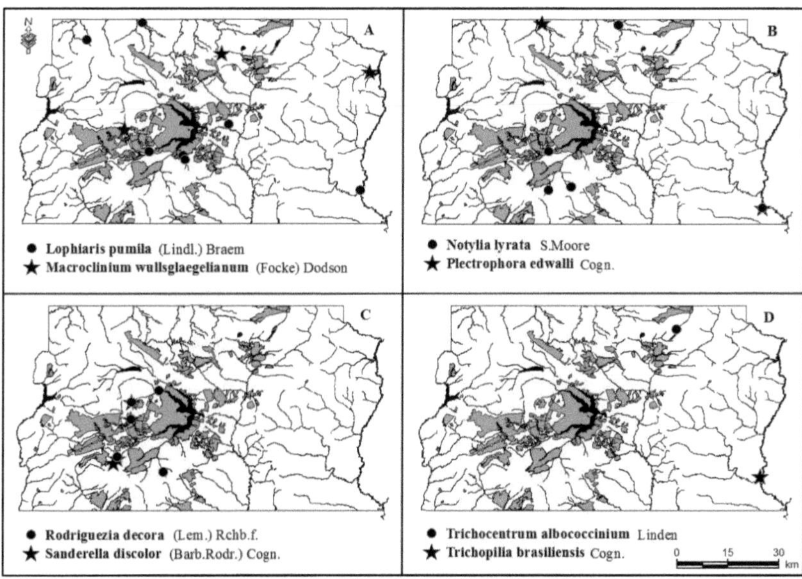

Figure 5 - Distribution of Oncidiinae species in the Federal District, Brazil.

49

7. BIBLIOGRAPHICAL REFERENCES

Aguiar, J.M.R.B.V. 2014. Reproductive biology of Ionopsis Kunth (Orchidaceae) from Brazil. **Master's thesis.** University of Sao Paulo. Ribeirao Preto. 63 p.

Aliscioni, S.S.; Torretta, J.P.; Bello, M.E.; Galati, B.G. 2009. Elaiophores in *Gomesa bifolia* (Sims) M.W. Chase & N.H. Williams (Oncidiinae: Cymbidieae: Orchidaceae): structure and oil secretion. **Annals of Botany** 14: 1141-1149.

Baptista, D.H.; Harding, P.A. & Neto, A.D. 2011. Orchids of Brazil: Oncidiinae I. **Associaçâo Orquidófila Piracicabana.** 1 edition. Brazil. P.: 21-29.

Barros, F. de & Batista, J.A.N. 2004. Varieties, forms and other infra-specific categories in Brazilian orchids. *In* : Fàbio de Barros & Gilberto Kerbauy (Org.), **Orquidologia sul-americana: uma compilaçâo cientifica.** p. 99-105. State Secretariat for the Environment, Botanical Institute, Sao Paulo - SP. 192 p.

Barros, F. de; Vinhos, F.; Rodrigues, V.T.; Barberena, F.F.V.A.; Fraga, C.N.; Pessoa, E.M.; Forster, W.; Menini Neto, L. 2015 Orchidaceae. *In*: **List of Species of the Flora of Brazil**. Rio de Janeiro Botanical Garden.

http://floradobrasil.jbrj.gov.br/jabot/floradobrasil/FB19976 <accessed: 24/02/2015>

Batista, J.A.N. & Bianchetti, L.B. 2003. Updated list of the Orchidaceae of the Federal District, Brazil. **Acta Botanica Brasilica** 17(2): 183-201.

Batista, J.A.N.; Bianchetti, L.B. & Pellizzaro, K.F. 2005. Orchidaceae Reserva Ecològica do Guarà, Distrito Federal, Brazil. **Acta Botanica Brasilica** 19(2): 221-232.

Bianchetti, L.B.; Batista, J.A.N.; Pellizzaro, K.F. & Augusto, M.M. 2005. Orchidaceae family in the Cafuringa APA. Chap. 4.6, pp: 153-163. *In*: Braga Netto, P.; Mecenas, V.V.; Cardoso, E.S. (Eds.), **APA de Cafuringa: a ùltima fronteira natural do DF**. Secretariat for the Environment and Water Resources. Brasilia: Semarh, 2005. 543 p. : il.

Braem, G.J., 1993. Studies in the Oncidiinae - discussion of some taxonomic problems with description of *Gudrunia* Braem, *gen. nov.* and reinstatement of the genus *Lophiaris* Rafinesque. **Schlechteriana** 1-2/93: 17-21.

Cameron, K.M., Chase, M.W., Whitten, M.W., Kores, P.J., Jarrel, D.C., Albert, V.A., Yukuwama, T., Hills, H.G., Goldman, D.H. 1999. A phylogenetic analysis of Orchidaceae: a evidence fron *rbcl* nucleotide sequences. **American Journal of Botany** 86: 208-224.

Carnevali, G.; Cetzal-Ix.; Balam, R.; Leopardi, C.;. Romero-Gonzalez, G.A. 2013. A combined evidence phylogenetic re-circumscription and a taxonomic revision of *Lophiarella* (Orchidaceae: Oncidiinae). **Systematic Botany** 38(1): 46-63.

Cetzal Ix, W.R.C.; Fernàndez-Concha, G.C.; Castro, V.P. 2012. *Cohniella* (Orchidaceae: Oncidiinae) South of the Amazon River. **Systematic Botany,** 37(1):58- 77.

Chase, M.W. and Palmer, J.D. 1989. Chloroplast DNA systematic of liliod monocots: Resources, feasibility, and an example from the Orchidaceae. **Amer. J. Bot.** 76:17201730.

Chase, M.W. 2009. Subtribe Oncidiinae. Pag. 211-391. *In:* Pridgeon, A.M.; Cribb, P.J.;

Chase, M.W. & Rasmussen, F.N. **Genera Orchidacearum. Epidendroideae (Part Two).** V. 5. Oxford. University Press.

Chase, M.W.; Williams, N.H.; Faria, A.D.; Neubig, K.M.; Amaral, M.C.E.; Whitten, M.W. 2009. Floral convergence in Oncidiinae (Cymbidieae; Orchidaceae): an expanded concept of *Gomesa* and a new genus *Nohawilliamsia*. **Annals of Botany** 104(3): 392.

Christenson, E.A. 1996. Notes on neotropical Orchidaceae II. **Lindleyana** 11(1): 12-26.

CRIA - Distributed Information System for Collections - Species Link - http://splink.cria.org.br// <access: 01/06/2014>

Docha Neto, A. & Benelli, A.P. 2006. *Alatiglossum culuenense*: a new species of Orchidaceae from Mato Grosso, Brazil. **Orchidstudium** 5:55-77.

Docha Neto, A. 2007. Taxonomic synopsis of the genus *Coppensia* Dumort: updated description and key to species. **Orchidstudium - International Journal of Orchid Study** 2(1): 14-22.

Dodson CH. 1962. The importance of pollination in the evolution of the orchids of tropical America. **American Orchid Society Bulletin** 31: 525-534, 641-649, 731-735.

Dodson C.H. 1965. Agents of pollination and their influence on the evolution of the Orchidacea family. **National University of the Peruvian Amazon.** Iquitos, Peru 1128 pp.

Dodson C.H. 2003. **Native Ecuadorian Orchids.** Volume 5. The Dodson Trust: Sarasota, Florida, USA.

Dodson, C.H. 2003. Why are there so many Orchid Species? **Lankesteriana** 7: 99-103. 2003.

Dressler, R.L. 1981. **The Orchids: Natural History and Classification.** Cambridge, Harvard University Press. P.: 1-154.

Dressler, R.L. & Williams, N.H. 1982. Proposal for the Conservation of the Generic Name 1779 *Oncidium* Swartz (Orchidaceae) with a Type Conserved Species, *Oncidium altissimum* Sw. **Taxon** 31: 752-754.

Dressler, R.L. 1993. **Phylogeny and classification of the orchid family.** Cambridge University Press.

Faria A. 2004. Phylogenetic systematics and delimitation of the genera of the subtribe Oncidiinae (Orchidaceae) endemic to Brazil: *Baptistonia, Gomesa, Ornithophora, Rodrigueziella, Rodrigueziopsis* and *Oncidium pro parte.* **PhD thesis.** State University of Campinas, Sao Paulo - SP, Brazil.

Felix L.P., Guerra M. 2000. Cytogenetics and cytotaxonomy of some Brazilian species of Cymbidioid orchids. **Genet Mol Biol.** 23:1-27.

Flora brasiliensis - the work. http://florabrasiliensis.cria.org.br//, <accessed: 02/01/2015>

Fernàndez-Concha, G.C., Cetzal Ix, W.R.C., Narvâez, R.B., & Romero-Gonzâlez, G.A. 2010. A synopsis of *Cohniella* (Orchidaceae, Oncidiinae). **Brittonia** 62(2), P.: 153-177.

Hoehne, F.C. **Iconografia de Orchidaceas do Brasil.** Secretary of Agriculture - Sao Paulo - Brazil, 1949. Pag 11-32

Garay, L.A. & Stacy, J.E. 1974. Synopsis of the Genus *Oncidium*. **Bradea** 40(1): 393424

INCT Virtual Herbarium of Flora and Fungi, 2014: <http://inct.florabrasil.net/> Accessed on: 15/12/2014.

Jiménez-Machorro, R. & Carnevali, G. 2001. Nomenclatural notes: new combinations in *Lophiaris* Raf. (Orchidaceae). **Harvard Papers in Botany** 6: 283-284.

Koniger, W. & Pongratz, D. 1997. *Stilifolium*: a new name for the section *Cebolletae* of the genus *Oncidium* as a new genus in subtribe Oncidiinae. **Arcula** 7: 186-190.

Koniger, W. & Pongratz, D. 1999. Intended combination not made: *Oncidium sprucei.* **Arcula** 9: 265.

Kranzlin, F. (1922). Orchidaceae-Monandrae: Tribus Oncidiinae-Odontoglosseae, 2. *In:* Engler, A. (ed) *Das Pflanzenreich.* Wilhelm Engelman, Leipzig, 344 pp.

Lemos, J.V.; Smidt, E.C.; Silva-Pereira, V. 2010. Floral visitors of *Gomesa recurva* R. Br. in mixed ombrophilous forest in Paranâ. **Book of Abstracts.** 61st National Botanical Congress. Manaus, Amazonas, Brazil.

List of Brazilian Flora Species. Rio de Janeiro Botanical Garden. Available at: <httpVfloradobrasil.jbrj.gov.br/>. Accessed on: 24/02/2015

Maury, C.M.; Ramos, A.E. & Oliveira, P.E. 1994. Floristic survey of the Aguas Emendadas ecological station. **Bulletin of the Ezechias Paulo Heringer Herbarium** 1: 46-67.

Mendonça, R.C.; Felfili, J.M.; Walter, B.M.T.; SilvaJunior, M.C.; Rezende, A.V.; Filgueiras, T.S.; Nogueira, P.E. & Fagg, C.W. 2008. Vascular flora of the Cerrado biome. In: Sano, S.M.; Almeida, S.P.; Ribeiro, J.F. (eds.). **Cerrado: ecology and flora**. Embrapa Informaçao Tecnològica.

Meneguzzo, T.E.C.; Bianchetti, L.B.; Proença, C.E.B. 2012. The genus *Encyclia* (Orchidaceae) in the Federal District, Goiàs and Tocantins. **Rodriguésia** 63(2): 277-292.

Menezes, L.C. 1995. *Rodriguezia decora* var. *lactea* L.C.Menezes var. nov. **Bol. CAOB** 5(1): 11.

Miller, D.; Warren, R.; Miller, I.M.; Seehawer, H. 2006. **Serra dos Órgaos, Its History and Its Orchids**. Scart, Rio de Janeiro, 574 p.

Neubig, K.M., Whitten, W.M., Williams, N.H., Blanco, M.A., Endara, L., Burleigh, J. G., Silvera, K., Cushman, J.C. & Chase. M.W. 2012. Generic recircumscriptions of Oncidiinae (Orchidaceae: Cymbidieae) based on maximum likelihood analysis of combined DNA datasets. **Botanical Journal of the Linnean Society** 168: 117-146.

Nogueira, P.E.; Nòbrega, M.G.G. & Pereira da Silva, G. 2002. Floristic survey and physiognomies of the Ezechias Heringer Ecological Park (Parque do Guarà) Federal District, Brazil. **Boletim do Herbârio Ezechias Paulo Heringer 10**: 31-56.

Pansarin, E.R. & Pansarin, L.M. 2011. Reproductive biology of *Trichocentrum pumilum*: an orchid pollinated by oil-collecting bees. **Plant Biology (Stuttgart):** 576581.

Pansarin, L.M.; Pansarin, E.R. & Santos, I.A. 2012. Comparative floral biology between two species of Oncidiinae (Orchidaceae): *Gomesa varicosa* M.W.Chase & N.H.Williams and *Comparettia coccinea* Lindl. **Book of Abstracts.** www.63cnbot.com.br (2012).

Parra-Tabla, V., Vargas, C.F., Magana-Rueda S., & Navarro J. 2000. Female and male pollination success of *Oncidium ascendens* Lindey (Orchidaceae) in two contrasting habitat patches: forest vs agricultural field. **Biological Conservation** 94: 335-340.

Pellizzaro, K.F.; Batista, J.A.N. & Bianchetti, L.B. 2004. The genus *Oncidium* Benth. (Orchidaceae) in the Federal District, Brazil. **Boletim do Herbàrio Ezechias Paulo Heringer** 14: 128-143.

Penha, T.L.L., Corrêa, A.M. & Catharino, E.L.M. 2011. Chromosome numbers in *Kleberiella* V.P. Castro & Cath. (Orchidaceae, Oncidiinae) and related genera. **Acta Botanica Brasilica** 25(2): 466-475.

Pereira, B.A.S.; Silva, M.A. & Mendonça, R.C. de. 1993. Orchidaceae. *In*: **Reserva Ecològica do IBGE, Brasilia (DF): lista das plantas vasculares**. IBGE, Rio de Janeiro. Pag. 35-36.

Pridgeon, A.M., Cribb, P.J., Chase, M.W. & Rasmussen, F.N. (eds.). 1999. **Genera**

Orchidacearum, v. 1. Oxford University Press, New York.

Pridgeon, A.M., Cribb, P.J., Chase, M.W. & Rasmussen, F.N. (eds.). 2009. **Genera Orchidacearum. Epidendroideae (Part Two)**. V. 5. Oxford. University Press.

Proença, C.E.B.; Munhoz, C.B.R.; Jorge, C.L.; Nóbrega, M.G.G. 2001. List and level of protection of the species of phanerogams of the Federal District, Brazil. *In* : T.B.

Cavalcanti & A.E. Ramos (Orgs.), **Flora of the Federal District, Brazil. Embrapa Recursos Genéticos e Biotecnologia** - Embrapa Cenargen, Brasilia - DF. p: 89-359.

Pupulin, F. 1995. A revision of the genus *Trichocentrum* (Orchidaceae: Oncidiinae). **Lindleyana** 10: 183-210.

Pupulin, F. & Carnevali, G. 2005. *Cohniella* Pfitz. Pp. 141-147. In: **F. Pupulin (ed.), Vanishing Beauty: Native Costa Rican Orchids** Vol. I. Costa Rica Univ. Press, San José, Costa Rica.

Rodrigues, V.T. 2011. **Orchidaceae juss. Morphological and taxonomic aspects.** Botanical Institute, Sao Paulo.

Roubik, D.W. & Ackerman, J.D. 1987. Long-term ecology of euglossine orchid-bees (Apidae: Euglossini) in Panama. **Oecologia** 73:321-333.

Salguero-Faria J.A.; Ackerman J.D. 1999. A nectar reward: Is more better? **Biotropica** 31: 303-311.

Sandoval-Zapotitla, E. & Terrazas, T. 2001. Leafanatomyof 16 taxa of the *Trichocentrum* clade (Orchidaceae: Oncidiinae). **Lindleyana** 16: 81- 93.

Sarmiento J. 2007. The Orchidaceae family in Colombia. Biological updates. **IV Colombian Congress of Botany.** Universidad de Antioquia, Medellin-Colombia.

Senghas, K. 1998. **Die Orchideen: Subtribus Oncidiinae.** Berlin: Parey.

Silva, D.G. 1999. The genus *Oncidium* Benth. (Orchidaceae) in the Chapada Diamantina, Bahia, Brazil. **Master's thesis.** State University of Feira de Santana, Feira de Santana - BA. 41 p. il.

Silvera, K. 2002. Adaptive radiation of oil-reward compounds among neotropical orchid species (Oncidiinae). Unpublished **M.S. thesis**, University of Florida.

Singer, R.B., & Cocucci, A.A. 1999. Pollination mechanisms in four sympatric southern Brazilian Epidendroideae orchids. **Lindleyana** 14, 47-56.

Singer, R.B.; Marsaioli, A.J.; Flach, A. & Reis, M.G. 2006. The ecology and chemistry of pollination in Brazilian orchids: recent advances. In: Teixeira da Silva J. (ed.) **Floriculture, ornamental and biotechonology: advances and topical issues.** Vol. IV. Isleworth: Global Science

Books, 570-583 p.

Sosa, V., Chase, M.W., Salazar, G.E., Whitten, W.M. & Williams, N.H. 2001. Phylogenetic position of *Dignathe* (Orchidaceae: Oncidiinae): evidence from nuclear ITS ribosomal DNA sequences. **Lindleyana** 16: 94-101.

Stafleu, F.A. 1985. Report of the Committee, for Spermatophyta: Proposal 679.1779 *Oncidium* Swartz (orchidaceae); *O. altissimum* Sw. typ. cons. prop. **Taxon** 34(4): 661.

Stpiczynska, M.; Davies, K.L.; Pacek-Bieniek, A.; Kaminska, M. 2013. Comparative anatomy of the floral elaiophore in representatives of the newly re-circumscribed *Gomesa* and *Oncidium* clades (Orchidaceae: Oncidiinae). **Annals of Botany** 112: 839854.

Szlachetko D.L., 1995. **Systema Orchidalium**. Fragm. Flor. Geobot., Suppl. 3: 1-152.

Torretta, J.P.; Gomiz, N.E.; Aliscioni, S.S.; Bello, M.E. 2011. Reproductive biology of *Gomesa bifolia* (Orchidaceae, Cymbidieae, Oncidinae). **Darwiniana** 49(1): 16-24.

Tropicos, 2014. Missouri Botanical Garden. http://www.tropicos.org// <accessed: 10/12/2014>

UNESCO. 2000. **Vegetation in the Federal District: Time and Space.** Brasilia. 74 p.

Van der Pijl, L. & Dodson, C.H. 1966. **Orchid Flowers: Their Pollination and Evolution.** University of Miami Press: Coral Gables, Florida, USA.

Walter, B.M.T. & Cavalcanti, T.B. 2005. **Fundamentals for collecting plant germplasm.** Brasilia, DF: Embrapa Recursos genéticos e Biotecnologia, 761 p.

Walter, B.M.T. & Sampaio, A.B. 1998. **The vegetation of the Sucupira farm**. Embrapa Recursos Genéticos e Biotecnologia, Brasilia.

Warming, E. 1973. **Lagoa Santa**. Belo Horizonte: Ed. Itatiaia; Sao Paulo: EDUSP, 1973. 284p. Original from 1892. Includes A vegetaçao de cerrados brasileiros, by M.G. FERRI.

Warming, E. 1892. **Lagoa Santa: Et Bidrag til den biologiske Plantegeografi.** Kgl. Danske Vidensk. Selsk. Skr., Raekke, naturvidensk. Or math. Afd 6(3). Kjobenhavn, Bianco Lunos Kgl. Hof-Bogtrykkeri.

WCSP (2014). 'World Checklist of Selected Plant Families. Facilitated by the Royal Botanic Gardens, Kew. Published on the Internet: http://apps.kew.org/wcsp/ <accessed: 01/06/2014>.

Weberling, F; Schwantes, H.O. 1986. Systematics of angiosperms. *In* : **Plant Taxonomy**. P.:129-131. Sao Paulo. EPU.

Williams, N.H. 1974. Taxonomy of the genus Aspasia Lindley (Orchidaceae: Oncidieae). **Brittonia** 26: 333-346.

Williams, N.H.; Chase, M.W.; Fulcher, T.; Whitten, W.M. 2001. Molecular systematics of the Oncidiinae based on evidence from four DNA sequence regions: expanded circumscriptions of *Cyrtochilum, Erycina, Otoglossum* and *Trichocentrum and* a new genus (Orchidaceae). **Lindleyana** 16: 113-139.

Whitten, W.M., Williams, N. H., & Chase, M.W. 2000. Subtribal and generic relationship of Maxillarieae (Orchidaceae) with emphasis on Stanhopeinae: combined molecular evidence. **American Journal of Botany** 87:1842-1856.

Zimmerman, J.K., Aide, T.M. 1987. Patterns of flower and fruit production in the orchid *Aspasiaprincipissa* Rchb.f. **American Journal of Botany** 74: 661-661.

ANNEX I

<u>DESCRIPTION **GENRES**</u>

- **herb**: grass

-       **behavior**: terrestrial, epiphyte, rupicola (**virgula,**)

-       **type of growth**: monopodial, sympodial (**dot.**)

* **rhizome**: appearance, interval between pseudobulbs (**dot.**)

- **secondary stems**:

a)   <u>swollen pseudobulbs</u> (**virgula,**)

- **pseudobulbs**: homoblastic or heteroblastic, presence or absence of sheaths and persistence of sheaths (**dot.**)

b)   <u>not swollen in pseudobulbs</u> (**comma,**)

- **leaves**: * presence or absence during flowering (**comma,**)

-       development (convoluted, conduplicated) (**comma,**)

-       forms (**comma,**)

-       disposition (apical and/or lateral in relation to the pseudobulb, or distal in relation to the stem) (**virgula,**)

-       articulated or not (**point.**)

- **inflorescence**: - position (terminal, lateral) (**virgula,**)

-       type (**comma,**)

-       whether or not protected by espata (**virgula,**)

-       approximate quantity and arrangement of flowers (dense, loose arrangements) (**virgula,**)

-       size in relation to the leaves (**dot.**)

- **flowers**: - resupinated or not (**comma,**)

- generalized size (large, small) (**comma,**)

- **sepals**: degree of similarity to the petals

a)   sepals and petals <u>similar to each other</u> (**virgula,**)

-       degree of adornment (**comma,**)

- some special feature (**point.**)

b)   different sepals and petals (**comma,**)

- size of the sepals in relation to the petals (**dot.**)

- **lip**: - general shape (whole, three-lobed) (**comma,**)

-      degree of attachment to the column (free, half-welded) (**comma,**)

-      **disc region**: - presence or absence of a callus (**comma,**)

- callus morphology (**dot**)

- **column**: - general form (**comma,**)

-      general characteristics: - winged; toothed apex; with auricles (**virgula,**)

-      basal extension (foot): present or not (**semicolon;**)

-      anther caducity (**semicolon;**)

-      number of lines, shape, consistency (**comma,**)

-      details of the pollinarium: (presence or absence of caudicle, viscidium) (**dot.**)

-      ovary: shape

-   Capsule: - general shape (dot.)

<u>DESCRIPTION **SPECIES**</u>

-   **behavior**: terrestrial, epiphyte, rupicola (**dot.**)

-   **rhizome**: appearance, interval between pseudobulbs

- secondary stems:

a)   swollen pseudobulbs

- **pseudobulbs**: shape, homoblastic or heteroblastic, color, size__________________

X __, presence or absence of sheaths and persistence, topped by ___________ leaves

(**point.**)

b)   not swollen pseudobulbs

- **stem**: shape, color, size __ X __ (**dot.**)

- **leaves**: * presence or absence during flowering (**comma,**)

-      number of leaves (**comma,**)

- sessile or *peciolate

*peciolode ____________ compr.

- general shape, base, margins, apex, consistency or texture (membranous, leathery), color, size X **(point.)**

- **inflorescence**: * type (when there is more than one type in the genus) **(comma,)**

- approximate number of flowers

- total size of the inflorescence (peduncle + rachis) **(semicolon;)**

- **flowers**: - showy, fragrant, small **(virgula,)**

- pedicellate or sessile **(virgula,)**

- length of pedicel + ovary **(point)**

- **sepals**: * degree of adnation **(comma,)**

* general shape, apex, margins, base, consistency or texture, color, dorsal sepal X , lateral sepals __ X ___ **(virgule,)**

* some special feature **(point.)**

* **petals**: - general shape, apex, margins, base, consistency or texture, color, size X **(comma,)**

* some special feature **(point.)**

- **lip**: - general shape (whole, three-lobed) **(comma,)**

- coloring

- size of the expanded lip (greatest width X greatest length) **(semicolon;)** or

- **lateral lobes**: - general shape **(virgula,)**

- form of presentation (position, patent, deflection) **(comma,)**

- color, size X_______________ **(semicolon;)**

* **median lobe**: - general shape **(virgula,)**

- some distinctive feature (emarginated apex, warty margin) **(virgula,)**

- presence or absence of isthmus **(comma,)**

- color, size X_______________ **(semicolon;)**

- **disc region**: * presence of callus, protruding veins **(virgula,)**

- callus morphology (**semicolon;**)

- **base**: anointed or not (**comma,**)

* nail size (**semicolon;**)

- **column**: - any special characteristics of the species, color, size

(**point.**)

Printed by Books on Demand GmbH, Norderstedt / Germany